Lecture Notes in Computer Scien

T0250703

Commenced Publication in 1973
Founding and Former Series Editors:
Gerhard Goos, Juris Hartmanis, and Jan van Leeuwen

Takeshi Okadome   Tatsuya Yamazaki
Mounir Makhtari (Eds.)

Pervasive Computing for Quality of Life Enhancement

5th International Conference On Smart Homes and
Health Telematics, ICOST 2007
Nara, Japan, June 21-23, 2007
Proceedings

 Springer

Volume Editors

Takeshi Okadome
NTT Communication Science Laboratories
Nippon Telegraph and Telephone Corporation
Japan
E-mail: houmi@idea.brl.ntt.co.jp

Tatsuya Yamazaki
Nataional Institute of Information and Communications Technology
Japan
E-mail: yamazaki@nict.go.ip

Mounir Makhtari
Groupe des Ecoles des Télécommunications
Institut National des Télécommunications
France
E-mail: Mounir.Mokhtari@int-edu.eu

Library of Congress Control Number: 2007928361

CR Subject Classification (1998): C.2.4, C.3, C.5.3, H.3-5, J.3-5, I.2, K.4, K.6

LNCS Sublibrary: SL 3 – Information Systems and Application, incl. Internet/Web
and HCI

ISSN   0302-9743
ISBN-10   3-540-73034-6 Springer Berlin Heidelberg New York
ISBN-13   978-3-540-73034-7 Springer Berlin Heidelberg New York

Springer is a part of Springer Science+Business Media

springer.com

© Springer-Verlag Berlin Heidelberg 2007

Typesetting: Camera-ready by author, data conversion by Scientific Publishing Services, Chennai, India
Printed on acid-free paper   SPIN: 12075720   06/3180   5 4 3 2 1 0

Preface

This volume of the LNCS series is a collection of papers presented in oral sessions at the Fifth International Conference on Smart Homes and Health Telematics (ICOST 2007) held in the ancient cultural city of Nara, Japan.

ICOST started in 2003 with the theme "Independent living for persons with disabilities and elderly people" and since then we have continuously provided a meeting place where researchers, practitioners, developers, policy makers and users meet to discuss a wide spectrum of research topics related to smart homes and services. Consequently, we feel that the participants of ICOST can share information on the state of the art, exchange and integrate different opinions and open up to collaboration.

Technologies are progressing persistently. Because of this progress, especially in the area of information and communication technologies, we believe that it is time to view the smart home technologies from a wider perspective and we have extended the focus of ICOST 2007 to "Pervasive Computing Perspectives for Quality of Life Enhancement." This theme has two meanings. One is how to construct pervasive computing environments as a common worldwide technology. The other is to examine in depth the definition of quality of life for different people.

Under the wide conference theme, we received 81 submissions. Through careful reviewing, we accepted 25 papers for oral presentations resulting in an acceptance rate of about 30%. Each paper was reviewed by at least two international reviewers and scored accordingly. Based on the scoring and all review comments, 25 papers were finally accepted to be included in this volume.

We would like to thank the Program Committee members for their hard work in reviewing papers within a very tight time schedule. We greatly appreciate all the Organizing Committee members who devoted their time in preparing and supporting this conference. Also, we would like to thank all the authors, the session chairs, the reviewers and all contributors to this conference who helped make it such a success.

June 2007   Tatsuya Yamazaki

Organization

We would like to acknowledge the support of the Nara Institute of Science and Technology (NAIST), the Commemorative Organization for the Japan World Exposition('70), the Support Center for Advanced Telecommunications Technology Research (SCAT), the Information Processing Society of Japan Kansai Branch, the Kao Foundation for Arts and Sciences, the Nara Convention Bureau, the Institute of Electronics, Information and Communication Engineers (IEICE), and the Research Promotion Council of Keihanna Info-Communication Open Laboratory.

Organizing Committee

Organizing Chair	Tatsuya Yamazaki (NICT, Japan)
Program Chair	Yasuo Tan (JAIST, Japan)
Publication Chair	Takeshi Okadome (NTT, Japan)
Finance Co-chairs	Nobuhiko Nishio (Ritsumeikan University, Japan)
	Motoyuki Ozeki (Kyoto University, Japan)
Local Arrangement Co-chairs	Masahide Nakamura (Kobe University, Japan)
	Etsuko Ueda (Nara Sangyo University, Japan)
Exhibition Chair	Kiyoshi Kogure (ATR, Japan)
Publicity Chair	Masaaki Iiyama (Kyoto University, Japan)
Conference Advisor	Masatsugu Kidode (NAIST, Japan)
Members	Akihiro Kobayashi (NICT, Japan)
	Takuya Maekawa (NTT, Japan)
	Goro Obinata (Nagoya University, Japan)
	Takuya Maekawa (NTT, Japan)
	Junji Satake (NICT, Japan)
	Yasushi Yagi (Osaka University, Japan)
	Yutaka Yanagisawa (NTT, Japan)

Scientific Committee

Chair	Sumi Helal (University of Florida, USA)
Members	Borhanuddin Mohd Ali (University of Putra, Malaysia)
	Juan Carlos Augusto (University of Ulster, UK)
	Nadjib Badache (University of Sciences and Technologies, Houari Boumediene, Algeria)
	Z. Zenn Bien (Korea Advanced Institute of Science and Technology, Korea)
	Diane Cook (Washington State University, USA)

Sajal Das (University of Texas, Arlington, USA)
Simon Dobson (UCD, Dublin, Ireland)
Sylvain Giroux (Université de Sherbrooke, Canada)
Jadwiga Indulska (University of Queensland, Australia)
Stephen S.Intille (MIT, USA)
Jay Lundell (Proactive Health Laboratory, Intel, USA)
Mounir Mokhtari (GET/INT Institut National des
    Telecommunications, France)
Chris Nugent (University of Ulster, UK)
Jose Piquer (University of Chile, Chile)
Toshiyo Tamura (Chiba University, Japan)
Daqing Zhang (Institute for Infocomm Research, Singapore)
Xingshe Zhou (Northwestern Polytechnical University, China)
Ismail Khalil Ibrahim (Johannes Kepler University, Austria)

Program Committee

Chair   Yasuo Tan (JAIST, Japan)
Members   Jit Biswas (Institute for Infocomm Research, Singapore)
          Carl Chang (Iowa State University, USA)
          Choonhwa Lee (Hanyang University, Korea)
          Heyoung Lee (Seoul National University of Technology, Korea)
          Paul McCullagh (University of Ulster, UK)
          Michael McTear (University of Ulster, UK)
          Kwang-Hyun Park (KAIST, Korea)
          Hélène Pigot (Université de Sherbrooke, Canada)
          Geoff West (Curtin University of Technology, Australia)
          Zhiwen Yu (Kyoto University, Japan)
          Keith Chan (Hong Kong Polytechnic University, China)
          Jianhua Ma (Hosei University, Japan)
          Suzanne Martin (University of Ulster, UK)
          Gilles Privat (France Telecom, France)
          Yasuyuki Shimada (Kumamoto National College of Technology,
              Japan)
          Koutarou Matsumoto (Nagoya University, Japan)
          Bessam AbdulRazak (University of Florida, USA)
          Nick Hine (Dundee University, UK)
          Abdallah M'hamed (GET/INT, France)
          Pierre-Yves Danet (France Telecom, France)
          Johan Plomp (VTT, Finland)
          Rose Marie Droes (VU University Medical Center, Netherlands)
          Aljosa Pasic (ATOS ORIGIN, Spain)
          Taketoshi Mori (Tokyo University, Japan)
          Min-Soo Hahn (DML, ICU, Korea)
          Woontack Woo (U-VR Lab, GIST, Korea)
          Xinggang Lin (Tsinghua University, China)

Katsunori Matsuoka (AIST, Japan)
Ian McClelland (Philips, Netherlands)
Bettina Berendt (Humboldt University, Germany)
Yoshinori Goto (NTT, Japan)
Tomoji Toriyama (ATR, Japan)
Deok Gyu Lee (ETRI, Korea)
Cristiano Paggetti (MEDEA, Italy)
Yoshifumi Nishida (AIST, Japan)

Sponsored by:

Nara Institute of Science and Technology (NAIST), Japan
The Commemorative Organization for the Japan World Exposition('70), Japan
The Support Center for Advanced Telecommunications Technology
    Research (SCAT), Japan
The Kao Foundation for Arts and Sciences, Japan
Nara Convention Bureau, Japan
Institut National des Télécommunications (INT), France
Groupe des Ecoles des Télécommunications (GET), France

In Cooperation with:

Information Processing Society of Japan Kansai Branch

Supported by:

The Institute of Electronics, Information and Communication Engineers
    (IEICE), Japan
Research Promotion Council of Keihanna Info-Communication Open
    Laboratory, Japan

Table of Contents

Toward a Desirable Relationship of Artificial Objects and the Elderly: From the Standpoint of Engineer and Psychologist Dialogue

Kotaro Matsumoto and Goro Obinata

EcoTopia Science Institute, Nagoya Univ., Furo-cho, Chikusa-ku, Nagoya-city, Japan
k-matsumoto@esi.nagoya-u.ac.jp

Abstract. In this paper a psychologist discusses the desirable relationship between the elderly and Artificial Objects based on some questions raised by an engineering researcher. Concretely, the paper discusses the differences between Objects, Machines, and Artificial Objects. Through this, it reveals the characteristic of Artificial Objects and presents the following four points that can be potentially problematic with respect to the relationship between such objects and the elderly. Moreover, these points should also be kept in mind by engineering researchers when developing Artificial Objects. 1) Artificial objects move autonomously, and therefore users can not take the role of initiator. That is, people must accommodate to artificial objects. 2) Artificial objects still only have a limited learning ability for creating relationships with their users, including the understanding of contexts and a shared history not unlike that between people. 3) Even with increased leaning ability as a result of increased efficacy and expanded operating capacity, which in turn leads to the creation of better relationships between artificial objects and persons, there are limitations to such relationships. 4) There is a need to point out the problem of initially focusing too heavily on the functions that artificial objects should have. Especially for the elderly, engineers should take time to consider the relationships of dependency that the objects they design will gradually foster over time.

Keywords: Artificial Objects, The elderly, Psychologist, Difference.

1 Introduction: The Day Prof. Obinata Asked Me Some Questions

One day, I was asked the following questions from Prof. Obinata, a researcher in engineering.

- *Why do people interact with "Artificial Objects" modeled on algorithms?*
- *Are there no (ethical) problems with respect to artificial objects becoming play companions for people?*
- *Is it not an emergent problem that artificial objects have started to look as if they have their own intentions, that they can monitor us, or that they can suggest things to us?*
- *In what directions can we take the design of artificial objects in response to the various needs and characteristics of the elderly?*

T. Okadome, T. Yamazaki, and M. Mokhtari (Eds.): ICOST 2007, LNCS 4541, pp. 1–8, 2007.
© Springer-Verlag Berlin Heidelberg 2007

- *For example, auto-mobile manufacturers are not liable for car accidents. On the other hand, in the case of robotics it is generally thought that the manufacturers will assume liability in the case of accidents. But where does this difference come from, as both types of accidents are basically the same, i.e., between persons and objects. How can we approach this issue, therefore, if we cannot designate a clear difference between the two?*
- *During the intervention of artificial objects, users may experience drawbacks or problems that would not qualify as accidents. For example, the more effective artificial objects accomplish work tasks the more the elderly may become inactive. To what extent should we take such possibilities into consideration?*

Presumable a lot of thoughtful engineers have questions such as these. In this paper I will try to respond to these questions because I am a psychologist who focuses on the relationship between the environment and the maintenance of the QOL of the elderly. But as these themes have yet to be explored thoroughly in research, it is impossible to respond to all of them at once. With this understanding, I wish to start exploring the answers to these questions. This paper, therefore, will attempt to reveal the characteristics and differences among objects, machines, and artificial objects.

First, I give a brief definition of objects, machines, and artificial objects.

Objects: Objects are the general objects around us, such as pens, coats, stones, and plants, etc.

Machines: Machines are devices and objects comprised of machinery, such as personal computers, cars, and vending machines, etc.

Artificial Objects: Artificial objects are machines that are either autonomous or make us believe that they are. More concretely, these comprise robotics. Here it is noteworthy to mention why the words "artificial objects" is used here and not AI (Artificial Intelligence). First, the concept of AI has various meanings, and it is not exactly clear to what it precisely refers. Second, it is not the theme of this paper to compare the intelligence of machines with that of humans. Rather, I intend to discuss the differences of objects and machines. Moreover, AI research tends to have the tacit purpose of recreating human intelligence. That is the chief difference between this investigation and AI research, as I wish to discuss whether artificial objects can serve to enrich a person's life, and if so, in what direction future progress should take.

In the next section, I discuss the differences between machines and objects, and the differences between artificial objects, machines, and objects.

2 Objects

2.1 Primitive Relationships Between Objects and Persons

When trying to understand the primitive relationships between objects and persons, one can begin with a discussion of "transitional objects." This term was introduced by Winnicott, a psycho-analyst who wrote about his vast clinical experience. Winnicott understood the relationship between mother and child as one of "dependency." For example, infants do not release their teddy bears or blankets because such things help to decrease their "separation anxiety" from their mothers on their developmental course of transition from "absolute dependency" (six months before) to "relative dependency" (about one year). The teddy bear or blanket then becomes the "transitional object" (Winnicott, 1974).

The phenomenon described by the term "transitional object" suggests that objects can function as substitutes for one's mother in terms of being deeply dependent on them. Accordingly, care must be taken to insure that such objects are not just random objects because not all objects carry the substitutive function of "transitional object" for infants.

As described above, "transitional objects" show the possibility to function as substitutes for persons. Next, I will discuss whether human cognition (and especially that of infants) is divided according to persons and objects. Johnson and Motion carried out an experiment to see if infants can discern between "direct faces" and "averted faces," that is, faces looking at them and faces looking away from them, respectively. Results of the experiment showed that two-month old infants gazed at the "direct face" for longer periods of time, while five-month old infants gazed at the "averted face" for longer periods of time. These results were interpreted to mean that two-month old infants prefer to look at "direct faces" while five-month old babies prefer to look at "averted faces." Moreover, as a general conclusion, their results suggested that even newborns can distinguish between persons and others (Johnson & Motion, 1991).

There is also the experimental paradigm known as the Turing Test, in which judges must determine whether things are persons or objects, similar to the abovementioned study. The Turing Test is famous as a measure of the ability to distinguishing between persons and machines/artificial objects. However, Searle has raised an objection to this paradigm and answered with his own thought experiment, known as the "Chinese Room." Subsequently, other researchers have raised their own objections to Searle's experiment, and therefore the argument about how to distinguish persons from other objects, including artificial objects, has yet to be settled (Russell & Norvig, 1997).

The above discussion about the primitive relationships between objects and persons reveals the following three points: 1) there exists dependency between objects and persons, 2) not just any object can become a substitute, and 3) there are differences between persons and objects, but these differences are still unclear.

2.2   Relationships Between Objects and the Elderly

"Personal objects can play an important role in maintaining personal identity in late life and may function as a distinctive language for the expression of identity and personal meaning" (Rubinstein, 1987). For example, elderly persons with dementia often show the phenomenon known as the "delusion of theft," which may emerge from a critical sense that the absence of objects (property) is a threat to their own sense of being. Further, it is easy to imagine that the identity of the elderly is in a state of diffusion during times of environment transit (e.g., when going from one's home to a nursing home), when they can become deprived of objects.

Next, I enumerate briefly some experimental research that has revealed the function of objects for the elderly.

1) Objects as a representation of the self or one's life: "Different kinds of possessions tend to have different meanings and referents in the lives of the subjects" (Sherman & Newman, 1977); "Significant objects that refer to the self" (Rubinstein, 1987).

2) Objects as opportunities or mediation to act-poiesis: "Significant objects of giving and receiving"(Rubinstein, 1987); "Significant objects of care" (Rubinstein, 1987); "The sensuousness of objects" (Rubinstein, 1987).

3) Objects for mediation with others: "Connections to others" (Rubinstein, 1987); "Relative to men, significantly more women had cherished possessions and were more likely to associate them with self-other relationships" (Wapner, Demick, & Redondo, 1990).

4) Objects as representations of the past: "Significant objects as representations of the past" (Rubinstein, 1987); "Possessions served the major functions of historical continuity, comfort, and sense of belonging" (Wapner, Demick, & Redondo, 1990); "They (cherished objects) seemed to serve as reconstructive symbols in the lives of the older persons. A significant positive relationship was found between memorabilia and mood" (Sherman, 1991).

5) Objects as foundations for spending daily life smoothly: In cases when elderly do not have any of the objects listed above in 1) to 4), the following results have been reported. "The lack of a cherished possession was associated with lower life satisfaction scores" (Sherman & Newman, 1977); "Significant objects as defenses against negative change and events" (Rubinstein, 1987); "Relative to those residents without cherished possessions, those with possessions were better adapted to the nursing home" (Wapner, Demick, & Redondo, 1990).

Rowles, Oswald, and Hunter (2003) and Sherman and Dacher (2005) have noted that cherished objects do not exist alone, that is, they suggest that it is necessary to broaden our view to "context of life" (e.g., home) when considering such objects. Objects comprise the very context of the daily lives of persons, and thus the meaning of such objects depends on the particular "context of life."

3   Differences Between Machines and Objects

3.1   Definition of Machine

First, I will enumerate the definition of machine briefly as follows.

1) Machines work within an algorithm constructed by the machine's designer. Machines need direct operation by their users, and they do not learn the characteristics of the user operations.

2) There is a basic principle structure of input and output, that is, behaviorism. Moreover, machines exist independent of context.

3) Machines are seen in terms of function and efficiency. Further, these qualities are used to compare machines to determine superiority.

4) Machines of superior function and efficiency are valuable, that is, there is value in the progress of technology.

3.2 Differences Between Machines and Objects: First Discussion

Let us now discuss the differences between machines and objects. I will cover the following three points: 1) Dependency, 2) Comparability, and 3) Context and History.

1) Dependency. Machines work within an algorithm constructed by their designers. The fact that only designers partake in the construction of machines problematic as it means that users are left with basically no such opportunity and can therefore only operate the machines. In other words, the identity of the user does not matter. To think more about objects in this sense, let us examine a writing pen. Pens can be used in a variety of ways to write all kinds of characters and letters. For most users, some pens write well, and some write poorly. Indeed, the relationship between persons and pens depends on the act of writing, and therefore each relationship is unique. On the other hand, let us consider the word processing function of personal computers as a representative of machines. Of course just about anyone can use such machines, but the ways that letters can be typed are always the same, or at most they are limited to a certain number of typesets for each user. This is why persons and machines are not in a relationship of dependency.

2) Comparability. Machines can make comparisons between other machines with respect to function and efficiency for determining superiority, and they can be substituted. However, a blanket that has become a "transitional object" can not be easily substituted because it has a unique touch and smell. Machines, on the other hand, can be replaced when they become broken, and new functions and high efficacy are considered to be positive changes.

3) Context and History. Machines operate on the basic principle of behaviorism, and they are independent of context. Being the newest is valuable, that is, not having any history is a valuable trait with respect to machines. Designers can construct algorithms that include context and history to a certain extent. However, machines are still unable to cope with the various contexts in which we cope in our daily lives. The ability to cope in this way has been called "dexterity" (Bernstein, 1996). Because machines work by inner algorithms designed previous to their operation, the various surrounding contexts and states of being of others have no meaning for machines. That is to say, the subject is isolated.

3.3 Differences Between Machines and Objects: Second Discussion

In the first discussion, I discussed the differences between machines and objects, focusing mainly 1) Dependency, 2) Comparability, and 3) Context and History. The above views, however, do not always apply to the elderly. Specifically, some elderly do not welcome the introduction of new machines and struggle with them. For example, let us here consider the example of a television remote control. There are many buttons to learn when first using a remote control, and many of them are hard to understand. The more natural it becomes to press the buttons that one needs, the more easily the buttons that are unnecessary go overlooked. In this way, the person becomes accustomed to the remote control.

The more a new remote controller is used, the more the unnecessary buttons go overlooked, as the user makes use of their experiences with the remote control and learn to press only the necessary buttons. In this case of learning to use a remote controller, naturally, therefore, the machine and person are in a state of dependence, the machine can not be compared to others, and the relationship between the machine and person depends on context and history. Essentially, using a remote control is an act of dexterity and goes beyond simple behaviorism. With respect to the elderly, this may be also understood as the remote controller not being able to change easily to fit the needs of the elderly. Here, I believe that machines and objects can therefore be divided on the following two points.

1) Results not open to change. Compared to objects, machines do not have open-ended results. That is, the things that are realized by machine are all the result of pre-decisions and are never indefinite.

2) It is person that changes. Both objects and machines can become accustomed to persons. Machines, however, work only within the algorithm constructed by their designer. Objects, on the other hand, such as leather jackets, can change. Therefore it is the person that changes during interactions with machines.

4   Differences Between Artificial Objects and Machines and Objects

First, I enumerate a definition of artificial objects briefly as follows.

1) Artificial objects can learn according to algorithms constructed by their designers.

2) Many artificial objects operate on the basic principle of behaviorism, but a few depend on context to a limited extent.

3) Artificial objects do not need direct operation by their users, and many of them operate autonomously.

Let us now discuss the differences between artificial objects and machines and objects. Here I will introduce the following three points: 1) Learning, 2) Moving (Body), and 3) Context and Autonomy.

1) Learning. Artificial objects are different from machines because artificial objects can learn in response to their being used even though they are limited by the range of their designer's construction. For example, the more frequently we use artificial objects, the more exact and precise the response becomes. That is there is the possibility for constructing original relationships between artificial objects and users.

2) Moving (Body). Originality is one of the most important characteristics of moving artificial objects. Artificial objects can appear to have an understanding of things when they autonomously move toward us. In the case of machines and objects, it is generally the person who moves. For example, if a car represents the machine, then the driver is the person. In the case of a talking vending machine, then the person as the moving person does not change. Artificial objects, however, move autonomously in much the same way as persons (or at least appear to do so). This is the difference between machines and objects.

Let us now discuss the type of robots designed for supporting rehabilitation that were created by Erikson, Mataric, and Winstein (2005). Their robots operate alongside persons engaged in rehabilitation, and they have the role of praising and encouraging each person's achievements and reprimanding those who are lazy. In the videos of their study, situations can be seen in which a lady at work repeatedly gets irritated by the robots near her. To me, this is actually a very natural consequence of such an environment.

3) Context and Autonomy. Most artificial objects must be pre-constructed to correspond to various contexts. There are many artificial objects, including many humanoid robots made to resemble machines, that look autonomous to us, but in fact they are just following the algorithms made by their designers. On the other hand, there are a few artificial objects, such as Brooks' Creature, that correspond to the current surrounding context, although the physical environments in which it can operate are limited. Of course the intelligence of artificial objects still can not realize the ability to handle and deal with any kind of situation like a person.

5   Conclusion

In this paper I attempted to reveal the problem areas related to the elderly having to accept the introduction of artificial objects (i.e., humanoid robots and welfare robots) into their lives through a discussion of the following three types of "Environmental Objects": Objects, Machines, and Artificial Objects.

Based on this discussion, it can be said that the chief characteristics of artificial objects are that they 1) move autonomously without direct operation by users, and that they are 2) able to optimally respond to a limited range of user behavior by learning the characteristics of that behavior through interactions with the users.

When compared to objects and machines, therefore, do these characteristics of artificial objects, therefore, not cause problems for the elderly when such objects are introduced into their daily lives?

The above discussion leads to the following conclusion:

1) Compared to objects and machines, artificial objects move autonomously and do not depend on the user, and therefore the user can not take the role of initiator. That is, persons must accommodate to artificial objects.

2) Artificial objects still have a limited learning ability for deepening the relationship of dependency with their users, including the understanding of contexts and a shared history not unlike that between persons.

3) Even with increased leaning ability as a result of increased efficacy and expanded operating capacity, which in turn leads to the creation of better relationships between artificial objects persons, there are limitations to such relationships, and in particular with respect to the elderly. Psychology has yet to identify the specific mechanisms of dexterity which allow one to ignore certain types of information (and forgetting memories at the same time) while picking up only that information which is meaning, as described by affordance theory.

4) There is a need to point out the problem of initially focusing too heavily on the functions that artificial objects should have. Especially in case of the elderly, who

characteristically have decreased physical ability and are fixed into a particular way of daily life, we can expect the sudden introduction of artificial objects with highly sophisticated features to cause confusion. Engineers should take time to consider the relationships of dependency that the objects they design will gradually foster over time.

Through the above conclusion, compared to objects and machines, part of the problem areas related to the introduction of artificial objects into the lives of the elderly was clarified. In the future, when it is possible to construct relationships of dependency such as those described above, we can expect there to be increased discussion about such relationships becoming over-dependent. Indeed, there are already examples of over-dependency with respect to persons and objects and machines (e.g., video games and the Internet). I believe, optimistically, that these are issues that should be taken into consideration only after it becomes possible to construct such dependency relationships between artificial objects and persons.

References

Bernstein, N.A.: On Dexterity and Its Development. Lawrence Erlbaum Associates, NJ (1996)

Erikson, J., Mataric, M.J., Winstein, C.J.: Hand-off Assistive Robotics for Post-Stroke Arm Rehabilitation. In: Proceedings of the 2005 IEEE/ 9h International Conference on Rehabilitation Robotics. Chicago (2005) http://robotics.usc.edu/

Johnson, M.H., Motion, J.: Biology and Cognitive Development: The Case of Face Recognition. Blackwell, US (1991)

Rowles, G.D., Oswald, F. Hunter, E.G.: Interior Living Environments in Old Age. Annual Review of gerontology and geriatrics. Wahl, H.W., Scheidt, R.J. (eds.) vol. 23, pp. 167–194. Springer, NY (2003)

Rubinstein, R.L.: The Significance of Personal Objects to Older People. Journal of Aging Studies 1(3), 225–238 (1987)

Russell, S., Norvig, P.: Artificial Intelligence: A Modern Approach. Prentice-Hall, NJ (1995)

Sherman, E., Newman, E.S.: The Meaning of Cherished Personal Possessions for the Elderly. INT'L Journal of Aging and Human Development 8(2) 181–192 (1977)

Sherman, E.: Reminiscentia: Cherished Objects as Memorabilia in Late-Life Reminiscence. INT'L Journal of Aging and Human Development 33(2) 89–100 (1991)

Sherman, E., Dacher, J.: Cherished Objects and the Home: Their Meaning and Roles in Late Life. In: Rowles, G.D., Chaudhury, H. (eds.) Home and Identity in Late Life, pp. 63–79. Springer, Berlin, Heidelberg, New York (2005)

Wapner, S., Demick, J., Redondo, J.P.: Cherished Possessions and Adaptation of Older People to Nursing Home. INT'L Journal of Aging and Human Development 31(3), 219–235 (1990)

Winnicott, D.W.: Playing and reality. Harmondsworth, Penguin (1974)

A Framework of Context-Aware Object Recognition for Smart Home*

Xinguo Yu[1], Bin Xu[2], Weimin Huang[1], Boon Fong Chew[1], and Junfeng Dai[2]

[1] Institute for Infocomm Research, 21 Heng Mui Keng Terrace, Singapore 119613
{xinguo,wmhuang,bfchew}@i2r.a-star.edu.sg
[2] Department of Electrical and Computer Engineering, National University of Singapore,
Singapore 117543
{u0305027,u0303322}@nus.edu.sg

Abstract. Services for smart home share a fundamental problem—object recognition, which is challenging because of complex background and appearance variation of object. In this paper we develop a framework of object recognition for smart home integrating SIFT (scale invariant feature transform) and context knowledge of home environment. The context knowledge includes the structure and settings of a smart home, knowledge of cameras, illumination, and location. We counteract sudden significant illumination change by trained support vector machine (SVM) and use the knowledge of home settings to define the region for multiple view registration of an object. Experiments show that the trained SVM can recognize and distinguish different illumination classes which significantly facilitate object recognition.

Keywords: Video Surveillance, Smart Home, Object Recognition, Context–aware, Illumination, Object Recognition, Support Vector Machine.

1 Introduction

There is an increasing interest in developing smart home worldwide because it can greatly enhance our life quality thanks to its services in taking care of our beloved, protecting our property, and aiding us when unwelcome events occur [3, 6, 9]. The examples of the services could be home cleaning, object localization, people activity monitoring [10]. One of the shared problems in developing smart home services is object recognition, which recognizes objects registered via camera (in the rest of this paper, object recognition means specific object recognition from video rather than object class recognition, which classifies objects into classes such as car, tree, people, etc). The three main challenges in object recognition for smart home are: (a) potential sudden illumination change; (b) possible high similarity of registered objects; (c) temporary occlusion of an object. Due to existence of case (c), it is impossible to track all registered objects all the time. For instance, when we monitor an object, it may be invisible due to occlusion or temporary lights-off; and when we re-see the object, it might have been moved to another place or replaced by a very similar object.

---

* This work is partially supported by EU project ASTRALS (FP6-IST-0028097).

T. Okadome, T. Yamazaki, and M. Mokhtari (Eds.): ICOST 2007, LNCS 4541, pp. 9–19, 2007.

Object recognition for smart home is a difficult application-oriented problem. Though object recognition can be reduced to a generic pattern classification problem, object recognition in a smart home context has its own uniqueness as listed below:

- The input of object recognition is video rather than image as our data input device is video camera.
- Object recognition is a part of monitoring and tracking physical objects in a smart home which makes it possible to learn object models through video.
- Some context knowledge such as settings of home, settings of camera, and illumination of the scene can be acquired to facilitate object recognition.

In object recognition field, a lot of techniques have been proposed such as textural features [13], illumination-invariant features [1, 4], geometric invariants [1], color-based methods [11], and local descriptors [2, 5, 7]. These techniques indicate great progress in object recognition; however more efforts are needed to solve the problem. The knowledge of context has various usages in different applications. It has been applied to computer network systems to provide context-aware services in computer [3]. In object recognition, it has been used to validate or infer object class [15, 16]. However, for object recognition the object models are not learned and organized according to context knowledge yet. In this paper, we *explicitly* integrate the knowledge of context into the framework for object recognition and show that it can enhance the performance of object recognition in real applications.

In the past decade, a bag of local descriptors were proposed and compared [2, 5, 7]. Scale Invariant Feature Transform (SIFT) proposed by Lowe [5] has been proven to outperform most other descriptors in image matching, object recognition, and object class recognition [7] (object class recognition means to recognize the classes of objects. SIFT is probably the most appealing descriptor for object recognition and object class recognition because it is scale-invariant, rotation-invariant, and partial illumination-invariant. Despite the proven capability for object recognition, SIFT is yet to be intensively tested in real application, at least to the best of our knowledge. In addition, SIFT descriptors cannot distinguish highly similar objects. For example, objects that differ only in size or color in a smart home context. It handles neither significant illumination change nor large-degree view change.

This paper develops a framework of object recognition for smart home which takes a context-aware approach, i.e. we extract the knowledge of context and use it to enhance the performance of object recognition. The framework uses SIFT as its key object recognition techniques, with the integration of the context knowledge to strengthen the recognition ability. By analyzing the color of fixed background, it recognizes illumination class of an image. The proposed framework has two phases; the first phase is to acquire the knowledge of context, including recognizing the illumination of the frames and finding possible candidate locations of the object; second phase uses model matching to recognize the object. The core idea is to integrate the knowledge of context acquired with SIFT descriptor for object recognition.

2   Problem Statement and Framework

Object recognition for smart home is a specific object recognition problem in a confined context, which can be divided into three sub-problems. The first problem is context knowledge acquisition; the second problem is object template acquisition and registration; the third problem is object recognition using acquired context knowledge and descriptor-based template matching. Thus, a complete system should comprise registration scheme, procedure of locating objects, and procedure of object model matching. The system requires an efficient registration of all objects. For good object recognition, several criteria should be satisfied. Firstly, it should tolerate illumination change; secondly, it should have strong distinguishing ability to differentiate from other similar objects.

Consequently, we have to address the following problems: (1) Object registration; (2) Illumination recognition; (3) Detection of candidate locations of objects; (4) Local object descriptor; and (5) Model matching strategy. These problems will be discussed in the following sections. Fig 1 is the block diagram of the framework of context-aware object recognition for smart home, which has two phases: (a) context knowledge extraction; and (b) object recognition based on context knowledge and local descriptor.

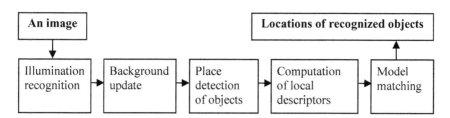

Fig. 1. Block diagram of the framework of context-aware object recognition for smart home

3   Object Registration

An object has to be represented properly for recognition because it appears differently in different illuminations, views, camera, and locations. We build a model for each situation of an object to holistically describe it. In this paper, an object model includes a representation of SIFT descriptors and other parameters such as dimension, color information, etc. Besides, there are a large number of models for each object considering the combination of location, view, camera, and illumination. To search the object model efficiently, we use a tree structure to organize the object models of all registered objects. This tree structure will help us to quickly reduce the number of models to be matched for identifying an object, using the features of context as "key" to search the tree. The hierarchy from top to bottom is Camera, Illumination, Location, Object, and View. In building this tree structure, there are two things to be considered: (1) how many illuminations, locations, and views are enough; (2) how we should identify or define them. Currently, we heuristically determine these parameters. For example, a surface of table is defined as a region and its center of region is defined as a location. For a wall, we divide it into several regions.

4 Illumination Recognition

Illumination is recognized using a trained SVM (Support Vector Machine) [14]. In the training phase, we extract features from the images with all sorts of illuminations to train the SVM. In the working phase, we extract features from the images and analyze the illumination. The features are color histogram, color difference histogram, contrast, and region brightness. Our experiment shows that these features can reliably reflect the class of illumination.

Color Histogram (Γ_1): The color histogram is the most basic and useful feature for many applications such as object detection. The most straightforward way to compute it is to accumulate all pixels in the given frame F. Normalization of this accumulative histogram yields the *average histogram*, which is the histogram used in this paper, denoted as Γ_1(F) or Γ_1.

4-Neighbour Color Difference Histogram (Γ_2): The absolute differences between each pixel and its four-neighbor (up, down, left, right) pixels are calculated and used to form the histogram. For each color channel of each pixel (i, j), we calculate a value

$$V_x(i,j)=|Y-X(i,j+1)|+|Y-X(i,j-1)|+|Y-X(i-1,j)|+|Y-X(i+1,j)|. \quad (1)$$

where $Y = X(i, j)$ and X means color channel, i.e. R or G or B.

Then we produce the histogram of these values for R, G, B separately.

Contrast (Γ_3): The contrast calculated for lighting conditions concentrates on the polarization. The kurtosis α_4 is used for this purpose: $\alpha_4 = \dfrac{\mu_4}{\sigma^4}$, in which μ_4 is the fourth moment about the mean and σ^2 is the variance. $\mu_4 = \sum_{i=1}^{N} (x_i - m)^4 \cdot p(x_i)$ and $\sigma^2 = \sum_{i=1}^{N} (x_i - m)^2 \cdot p(x_i)$, where x_i is the i-th pixel of the frame; N is the total pixel number; m is the mean value of the frame; $p(x_i)$ is the probability density function (PDF) for the i-th pixel. The final measure of contrast is as follows:

$$F_{contrast} = \sigma / (\alpha_4)^2 \quad (2)$$

Region Brightness (Γ_4): To gain more local variance information, a region based brightness feature is used. Each frame is divided into certain number of smaller regions. The brightness for each region is calculated and the values are stored in a new histogram for future use. For each region, brightness is the average scale values of each RGB channel.

$$B_r = \frac{1}{N} \cdot \sum_{i=1}^{N} x_i \quad (3)$$

where N is the total pixel number, x_i is the color value for the ith pixel.

5 Background Update and Location Candidate

Background update has to cope with two classes of illumination changes: sudden and gradual changes. For sudden change, the illumination recognition can detect its occurrence and we use the generic background of the detected illumination model as the background of the scene. For gradual change, we model the scene using a mixture of K Gaussians for each of the pixel locations, as Stauffer and Grimson did in [12]. The probability of a pixel having an intensity value X_t (n dimension vector) at time t can be written as:

$$P(X_t) = \sum_{i=1}^{K} \omega_{i,t} * \eta(X_t, \mu_{i,t}, \Sigma_{i,t}). \tag{4}$$

where $\eta(X, \mu, \Sigma) = \dfrac{1}{(2\pi)^{\frac{n}{2}} |\Sigma|^{\frac{1}{2}}} e^{-\frac{1}{2}(X_t - \mu_t)^T \Sigma^{-1}(X_t - \mu_t)}$,

$\omega_{i,t} = (1-\alpha)\omega_{i,t-1} + \alpha * M_{k,t}$, μ is mean, α is learning rate and $M_{k,t}$ is 1 for the model which is matched and 0 for other models, assuming that R, G, B are independent and have the same variances, $\Sigma_{k,t} = (\sigma_k)^2 I$. Then, the first H Gaussians are chosen as the background model, where

$$H = \arg\min_h (\sum_{k=1}^{h} \omega_{k,t} > T). \tag{5}$$

where T is minimum value of area ratio of background to whole image, assuming foreground objects only occupies a small area of the image.

Once we obtain the background of scene, we use background subtraction to identify the candidate locations of moved objects. Note we do not do object segmentation but we estimate locations of objects.

6 SIFT Descriptor

Lowe [5] presented a method to extract distinctive invariant features from images named as SIFT (Scale Invariant Feature Transform). SIFT descriptors are invariant to image scale and rotation. They are also proven to be partial invariant to affine distortion and illumination.

SIFT descriptors are well localized. The partial occlusion of an object only affects part of the descriptors thus we may still be able to recognize the object. Another important characteristic is that SIFT descriptors are very distinctive. The cost of extracting descriptors is minimized by taking a multiple step approach, in which the more expensive operations are applied only to locations that pass the previous steps. SIFT descriptors take the following four steps to calculate: Step 1, Scale-space extrema detection; Step 2, Accurate keypoint localization; Step 3, Orientation assignment; and Step 4, SIFT descriptor.

Because of good properties of SIFT descriptor, we use SIFT descriptor as main elements of our object model.

7   Model Matching Strategy

To recognize an object, the SIFT features computed for the observed object should be matched with the features of the object models learned. In practice, we can build a set of object models for each object. We can match them efficiently if the number of object models for an object is small. However, the accuracy of object recognition will drop. To resolve this issue, we propose a tree structure to organize the object models as described in section 2. In matching, we use two methods to speed up the matching procedure: (1) Number reduction for costly template matching and (2) Locating object at the previous location first. Before matching, we have known the illumination and places of objects. As such, we use the object models related with the illumination recognized and place detected to do matching. As the object models are organized in a tree structure, we can identify these object models very fast. In a smart home, registered objects should have very small chance to have their locations changed simultaneously. Hence, we can first locate objects at their previous locations and we should be able to locate most of the objects. For the remaining objects, we identify them at the remaining candidate locations.

8   Experimental Result

We test our algorithm on videos taken from a smart home built in Singapore.[1] Our test setting was two boxes of similar design pattern with nine classes of illuminations. The results on illumination recognition and object recognition are presented below. We also test on one book with plain surface to show the limitation of SIFT descriptor, explaining reasons why we need other features besides SIFT descriptors.

8.1   Results on Illumination Recognition

Nine original video clips with different lighting conditions are used for the experiment. Each clip corresponds to a class of basic illumination. Each frame is flipped horizontally to create classes 10 to 18 and then vertically to create classes 19 to 27. Therefore, there are 27 classes in total for the classification testing. Representative frames (one from each basic class) are shown in Fig 2. Training and testing data sets are separately extracted from the 27 classes with all the different combinations of the 4 features. The extracted data is trained and tested with SVM Torch [14]. The testing results are shown in the 4 graphs in Fig 3 with respect to each feature. Fig 3 shows that the 3 features (color histogram Γ_1, color difference histogram Γ_2, and contrast Γ_3), even they are combined, gave considerably high matching error ratio. This illustrates that the 3 features lack the ability to distinguish the local area changes of the lighting conditions. Fig 3 shows that region-brightness Γ_4 is the most effective feature. In fact, the results shows that region brightness alone can distinct the illuminations in our test data set. Of course, this may not hold for a larger data set so that for reliability we would not use region brightness alone to distinct the illumination classes.

---

[1] StarHome: http://starhome.i2r.a-star.edu.sg/

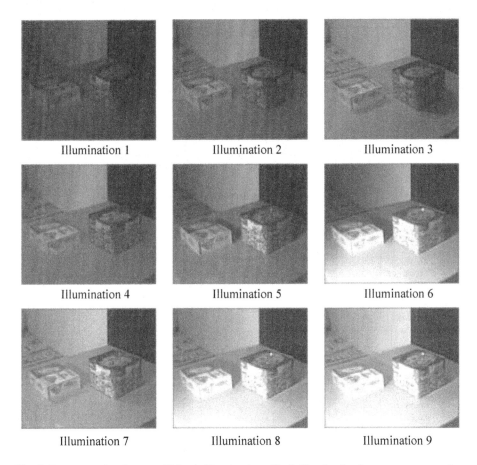

Fig. 2. Representative frames of 9 basic illuminations. Each illumination has one representative frame.

Table 1. Matrix of cross matching performance of nine templates and nine illuminations for one frame (Tested frames from nine illuminations I1-I9 for A and B two objects)

	T1		T2		T3		T4		T5		T6		T7		T8		T9	
	A	B	A	B	A	B	A	B	A	B	A	B	A	B	A	B	A	B
#D	61	37	81	71	72	68	99	83	95	74	106	76	135	69	125	75	129	61
I1	30	15	2	2	10	0	4	2	1	3	0	1	1	4	0	1	0	5
I2	3	3	41	36	7	14	16	13	10	8	2	3	6	9	1	4	4	10
I3	10	5	8	16	45	41	17	17	4	8	1	4	8	11	0	6	1	3
I4	5	5	15	13	16	14	62	37	11	9	2	5	21	14	2	4	4	7
I5	3	5	6	6	4	4	15	5	59	42	8	5	26	12	7	5	4	8
I6	2	3	2	3	1	4	1	4	7	5	82	42	5	5	35	18	30	14
I7	3	3	9	10	11	8	21	12	27	9	8	6	97	40	6	7	7	6
I8	0	2	5	6	0	8	3	4	7	7	43	18	6	9	84	44	37	19
I9	0	3	4	7	1	4	2	4	5	8	29	13	6	7	35	15	93	43

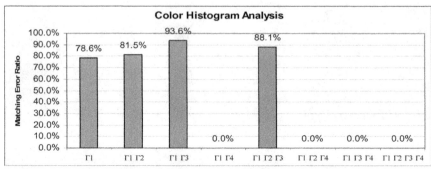

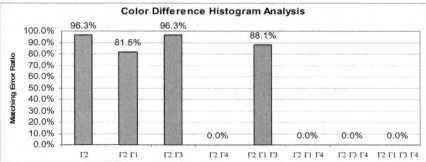

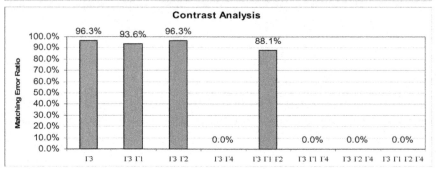

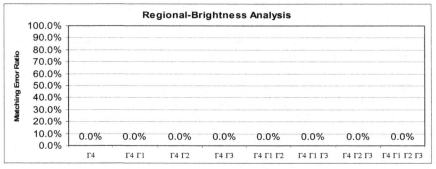

Fig. 3. Cross analysis among four features on illumination recognition

8.2 Object Recognition

For object recognition, our test focuses on how illumination recognition facilitates object recognition. The experimental setting is as follow: Two boxes, A and B with similar surface patterns are used. Nine illuminations are taken in the test. We build a SIFT template for each object in each class of illumination. One frame is arbitrarily selected from a video clip recorded in one illumination and thus nine frames are obtained, given in Fig 2. Then we use each template to match each frame to identify the object. Samples of template matching are given in Fig 4. The matrix of matching performance is given in Table 1-2. In Table 1, #D is the number of descriptors in frame; the number in the cell in column Ti (A or B) and row Ij is the number of matched descriptors between the template of object A or B in illuminate i and the frame from illumination j. Table 2 is similar to Table 1. However, the number in its

Table 2. Matrix of cross matching performance of nine templates and nine illuminations for 50 frames (Tested frames from nine illuminations I1-I9 for A and B two objects)

	T1		T2		T3		T4		T5		T6		T7		T8		T9	
	A	B	A	B	A	B	A	B	A	B	A	B	A	B	A	B	A	B
#D	61	37	81	71	72	68	99	83	95	74	106	76	135	69	125	75	129	61
I1	**49.2**	**40.5**	2.47	2.82	13.9	0.00	4.04	2.41	1.05	4.05	0.00	1.32	0.74	5.80	0.00	1.33	0.00	8.20
I2	4.92	8.11	**50.6**	**50.7**	9.72	20.6	16.2	15.7	10.8	10.8	1.89	3.95	4.44	13.0	0.80	5.33	3.10	16.4
I3	16.4	13.5	9.88	22.5	**62.5**	**60.3**	17.2	20.5	4.21	10.8	0.94	5.26	5.93	15.9	0.00	8.00	0.78	4.92
I4	8.20	13.5	18.5	18.3	22.2	20.6	**62.6**	**44.6**	11.6	12.2	1.89	6.58	15.6	20.3	1.60	5.33	3.10	11.5
I5	4.92	13.5	7.41	8.45	5.56	5.88	15.2	6.02	**62.1**	**56.8**	7.55	6.58	19.3	17.4	5.60	6.67	3.10	13.1
I6	3.28	8.11	2.47	4.23	1.39	5.88	1.01	4.82	7.37	6.76	**77.4**	**55.3**	3.70	7.25	28.0	24.0	23.3	23.0
I7	4.92	8.11	11.1	14.1	15.3	11.8	21.2	14.5	28.4	12.2	7.55	7.89	**71.9**	**58.0**	4.80	9.33	5.43	9.84
I8	0.00	5.41	6.17	8.45	0.00	11.8	3.03	4.82	7.37	9.46	40.6	23.7	4.44	13.0	**67.2**	**58.7**	28.7	31.2
I9	0.00	8.11	4.94	9.86	1.39	5.88	2.02	4.82	5.26	10.8	27.4	17.1	4.44	10.1	28.0	20.0	**72.1**	**70.5**

| (a) big box SIFT matching in the same illumination. | (b) big box SIFT matching in the different illumination. | (c) small box SIFT matching in the same illumination. | (d) small box SIFT matching in the different illumination. |

Fig. 4. Sample results of SIFT template matching. In figure, each green line represents a pair matched descriptors between template and image. In (a) to (d), the small pictures in upper part is the training image from which the templates are extracted; the large pictures in the lower part are the testing images. The corresponding templates are denoted as Ta, Tb, Tc, and Td; and the corresponding images are denoted as Ia, Ib, Ic, and Id. Ta and Tb (Tc and Td) are the same image. Ia and Ib (Ic and Id) are captured in two similar illuminations by human perception.

each cell gives the average percentage of matched descriptors with respective to #D for 50 frames. Fig 4 and Table 1-2 show that the matched numbers are significantly larger when i and j are the same number. This shows that illumination recognition can speed up the procedure of object recognition by matching only the templates in the recognized illumination class.

8.3 Limitation of SIFT Descriptor

SIFT descriptor has its limitations, though it is very distinctive for object recognition in general. One of limitations is that there are few SIFT descriptors for objects with plain surface such as an object in Fig 5 under three illuminations. It is obvious that we can use the color histogram to recognize the object if we built its template histogram under classes of illuminations.

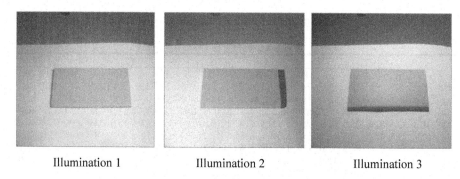

Illumination 1   Illumination 2   Illumination 3

Fig. 5. A book with plain surface under three illuminations

9   Conclusions and Future Work

We have presented a framework of object recognition for smart home that has addressed the issue of significant illumination change while monitoring the object of interest (or registered). As object recognition in smart home needs to distinguish the objects that may be very similar, we cannot use SIFT descriptor only. We proposed to use object model with not only SIFT descriptor playing the core role but other features such as color histogram in Fig 5 are also as important features. We have designed a tree structure to organize the models of all registered objects in a smart home. It can help us to narrow down to the branch of the tree for model matching, provided that we have obtained the context information. The experimental results show that the trained support vector machine can reliably recognize illumination condition. More importantly, the recognized illumination condition can greatly speed up the procedure of object recognition and improve the performance of object recognition.

There are still plenty of jobs to be done. First, we will improve the current algorithm to include more techniques of object recognition such as shape-based object recognition techniques. We will modify the learning process to be automatically done for the illumination class and object models of the registered object. Second, we want to study indirect object recognition techniques similar to what Persrseum did in [9].

Such techniques can help us to extract more information on the object. For example, provided that we have obtained the information of activity of human beings, we can further distinguish whether the object is occluded or removed. Finally, we want to integrate object recognition into larger home intelligent system. Another important aspect we want to explore is how to build the tree structure of models of objects. There are two main problems. The first problem is how to use less manual work to acquire these models of many views. The second problem is how we can use less memory to store the tree of large number of models.

References

[1] Alferez, R., Wang, Y.-F.: Geometric and illumination invariants for object recognition. IEEE Transactions on Pattern Analysis and Machine Intelligence, vol. 21(6), pp. 505–536 (1996)

[2] Abdel-Hakim, A.E., Farag, A.A., CSIFT,: A SIFT descriptor with color invariant characteristics. Computer Vision and Pattern Recognition 2, 1978–1983 (2006)

[3] Gu, T., Pung, H.K., Zhang, D.Q.: A service-oriented middleware for building context-aware services. Journal of Network and Computer Applications 28(1), 1–18 (2005)

[4] Liu, Q., Sun, M., Sclabassi, R.J.: Illumination-invariant change detection model for patient monitoring video. 26th Annual Int'l Conf of Eng. in Medicine and Biology Society 3, 1782–1785 (2004)

[5] Lowe, D.: Distinctive image features from scale-invariant keypoints. International Journal of Computer Vision 60(2), 91–110 (2004)

[6] Luo, R.C., Lin, S.Y., Su, K.L.: A multiagent multisensor based security system for intelligent building. IEEE conf on Multisensor Fusion and Intelligent Systems, pp. 2394–2399 (2003)

[7] Mikolajczyk, K., Schmid, C.: A performance evaluation of local descriptors. IEEE Transactions on Pattern Analysis and Machine Intelligence 27(10), 1615–1630 (2005)

[8] Mester, R., Aach, T., Dümbgen, L.: Illumination-invariant change detection using a statistical colinearity. In: Pattern Recognition. LNCS, vol. 2191, Springer Verlag, Heidelberg (2001)

[9] Mihailidis, A., Carmichael, B., Boger, J.: The use of computer vision in an intelligent environment to support aging-in-place, safety, and independence in the home. IEEE Trans. Inf. Technol. Biomed. 8(3), 238–247 (2004)

[10] Peursum, P., West, G., Venkatesh, S.: Combining image regions and human activity for indirect object recognition in indoor wide-angle views. ICCV 2005 1, 82–89 (2005)

[11] Seinstra, F.J., Geusebroek, J.M.: Color-based object recognition on a grid. ECCV 2006 (2006)

[12] Stauffer, C., Grimson, W.E.L.: Adaptive background mixture models for real-time tracking. CVPR99 (June 1999)

[13] Tamura, H., Mori, S., Yamawaki, T.: Textural features corresponding to visual perception. IEEE Trans. on Systems, Man, and Cybernetics SMC-8(6), 460–473 (1978)

[14] Collobert, R., Bengio, S.: SVMTorch: Support vector machines for large-scale regression problems. Journal of Machine Learning Research 1, 143–160 (2001)

[15] Torralba, A., Murphy, K.P., Freeman, W.T., Rubin, M.A: Context-based vision system for place and object recognition. ICCV 2003 1, 273–280 (2003)

[16] Strat, T.M., Fischler, M.A.: Context-based vision: recognizing objects using information from both 2D and 3D imagery. IEEE Transactions on Pattern Analysis and Machine Intelligence 13(10), 1050–1065 (1991)

A Pervasive Watch-Over System Based on Device, User and Social Awareness

Kazuhiro Yamanaka, Yoshikazu Tokairin, Hideyuki Takahashi,
Takuo Suganuma, and Norio Shiratori

RIEC/GSIS, Tohoku University
2-1-1 Katahira, Aoba-ku, Sendai, 980-5877, Japan
{yamanaka,tokairin,hideyuki,suganuma,norio}@shiratori.riec.tohoku.ac.jp

Abstract. In this paper we propose a gentle system for watching over for pervasive care-support services that fulfill users' actual requirements and without interfering the users' privacy. In this system, we cope with not only the user's physical location but also all the situations of pervasive devices and requirements of associated users, in "both" sides of watching and watched sites. Moreover, social relationships between observers and a watched person are used to give an adequate privacy level to the services. To do this, we propose a pervasive watching over system "uEyes". In uEyes we introduce three distinguished features for the watching, they are (S1) Device awareness, (S2) User awareness and (S3) Social awareness. To realize these features, an autonomous decision making ability and cooperative behavior are employed to each element in the system. Based on these advanced features, a live video streaming system is autonomously constructed according to the device status and the user's situation, on both sides in runtime. As a result, the gentle watching-over service can be effectively provided.

1   Introduction

Recently, there are several challenging works concerning pervasive computing environments for daily life support [1,2]. In this context, multimedia watching systems are widespread as one of the care-support systems that watches over children and elderly people from remote site connected by wide-area network [3,4,5]. Researches are at their initial stage to apply this system to pervasive computing environments. There are some researches on delivering video streaming from a selected camera mainly closest to the target person, by using physical positional information of the target [6,7]. Also, there exist works on flexible displaying that plays the video seamlessly from a nearest display to the observer based on the location of the watcher [8,9].

However, in these existing works, when "a watching person wants to see the facial color of the watched person's face with very high resolution, without violating his/her privacy", it is expected to satisfy such detailed users' requirements. To fulfill the above requirement, an autonomous system construction mechanism is required, according to not only the user's location, but also device statuses,

T. Okadome, T. Yamazaki, and M. Mokhtari (Eds.): ICOST 2007, LNCS 4541, pp. 20–30, 2007.

network availability, users' requests on the services and privacy concerns, in both the sides.

In this paper, we propose a scheme to cope with not only the user's physical location but also all the situations of pervasive devices and requirements of associated users, in "both" sides of watching and watched sites. Moreover, social relationships between watching people and a watched person are also used to give an adequate privacy level to the services. To realize this, we propose a ubiquitous watching over system "uEyes", to address the above issues. The original idea of uEyes can be described in following three points: (S1) Device awareness, (S2) User awareness and (S3) Social awareness.

We design and implement the features of uEyes by introducing an autonomous decision making ability and cooperative behavior to each system element, based on an agent-based computing framework AMUSE [11,12]. Employing the advanced features, a live video streaming system is autonomously constructed according to the individual context of the devices and the users on both sides in runtime, by cooperating among them. As a results, a gentle watching-over service that fulfils the detailed users' requirements can be effectively provided.

We implemented a prototype of uEyes for watching over elderly people, and performed some experiments based on several scenarios of watching by his/her family. We assumed a scenario, where a son is taking care of his elderly father who has heart disease, wanted to watch his facial color and expression in a high quality video. Then, the live video streaming system involving a high resolution camera and display devices was dynamically configured by uEyes. Also in case of his father in emergency, the privacy level was lowered and the situation was informed to members of the local community. From results of these case studies, it was confirmed that uEyes can provide real-time multimedia watching services for elderly people, with the reasonable QoS and privacy that meets the users' requirements.

2   uEyes: The Concept

Our target watching task is as shown in Fig.1(a). In this figure, a community of watching people cooperatively watches over a target person. We call this kind of watching task as a "Community-based watching-over" task. The purpose of uEyes is to solve the problem in this watching-over task. To achieve this, it is necessary to introduce a mechanism where entire system configuration that consists of various system elements is dynamically selected and organized according to the states of the system, users, situations around the users and social aspects, as shown in Fig.1(b). To realize this, following three characteristics are newly introduced:

(S1) *Device awareness:* uEyes effectively handles and coordinates multiple contexts of pervasive devices for provisioning of appropriate QoS of watching systems. The contexts involve not only the user location, but also the status of display/camera devices, available resources of PCs and hand-held devices, available network access and bandwidth, etc.

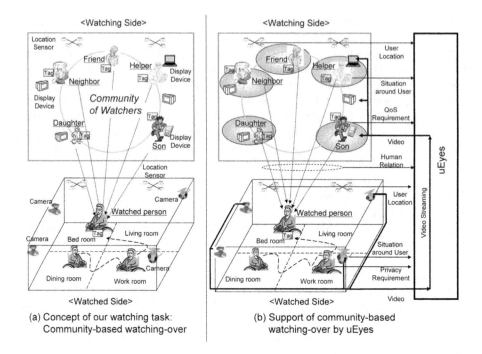

Fig. 1. Concept of uEyes

(S2) *User awareness:* uEyes closely associates with the user's requirement in the best possible way. For example, in case where a watcher requires the video streaming so that he can vividly view the facial color of the watched person, a high quality and zoomed picture should be appeared in the nearest display.

(S3) *Social awareness:* uEyes deeply considers the social relationship and keeps the adequate privacy according to the situation. For instance, in case of a normal situation, the watched person's privacy should be protected, however in case of emergency, the privacy level would moderately be lowered.

3   Design of uEyes

3.1   AMUSE Framework

We employ a multiagent-based framework for service provisioning in ubiquitous computing environments, called AMUSE (Agent-based Middleware for Ubiquitous Service Environment) [11,12], as a software platform to build uEyes. The basic idea of this framework is "agentification" of all the entities in the pervasive computing environments. Here, the agentification is a process to make a target entity workable and context-aware. The details are discussed in [11,12], so we omit them in this paper.

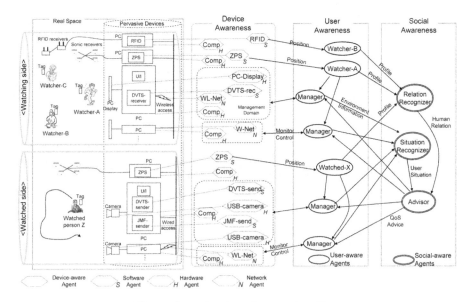

Fig. 2. Organization of agents in uEyes

In AMUSE, agents can perform the following advanced cooperation and intelligent behavior for the pervasive computing:

(1) Recognition of statuses of individual entities
(2) Coordination of contexts of multiple entities
(3) Service construction by combinations of entities
(4) Intelligent processing of the situations around the entities

3.2 AMUSE Agents for uEyes

Fig.2 shows agents in uEyes. The agents consist of Device-aware agents, User-aware agents and Social-aware agents. The agents basically reside in computers, and they manage corresponding entities that are connected to, or are running on the computer.

The Device-aware and User-aware agents cooperatively work to accomplish QoS that meets to user's requirements on a watching task and device situations. Multiple contexts of the devices should be deeply considered to achieve our goal. These contexts are individually maintained by each agent, and its effective coordination would be performed by cooperation among related agents. This cooperative behavior is performed by the following steps:

[Step-1: IAR updating]. Agents that are physically located closely exchange information on each context, and update an Inter-agent Relationship (IAR). The IAR represents a relationship that an agent has against another agent. For example, agents in the same PC potentially have tight relationship each other. Moreover,

the IAR for cooperation is updated according to histories of cooperative works between agents.

[*Step-2: User information acquisition*]. A User agent (we abbreviate "agent" to "Ag" hereafter) in the closest PC or the hand-held device collects user's information such as requirements and profiles. This information is employed to check whether provided QoS of the services are satisfied or not.

[*Step-3: Agent organization decision in initiate site*]. When the user moves into a sensor area, RFID Ag or ZPS Ag captures the movement and informs related agents based on the user's location. Then, only the related agents begin negotiation to make an agent organization based on the IAR. Here, the agents that achieved a good performance in past cooperation, are preferentially selected using the IAR.

[*Step-4: Agent organization decision in opposite site*]. It should be considered that agent organizations of both the sender and receiver sites work properly to accomplish the QoS. If privacy and emergency concerns are required, Social-aware agents are activated in this step.

[*Step-5: Service provisioning*]. After the agent organization is configured, live video streaming will start using the appropriate camera, software, access network, display, etc.

[*Step-6: User's feedback*]. After the end of the service, user's feedback is collected by the User Ag, and then the IAR is updated based on the user's evaluation.

3.3   Agent Behavior for Social Awareness

In this section, we describe a process of QoS control that considers privacy and emergency. This process is based on recognition of human relationships and environmental situations by Social-aware agents. In the initial step, Relation Recognizer Ag infers human relationships between an observer and a watched person using their profiles that each User Ag maintains. Relation Recognizer Ag utilizes ontology that represents background knowledge to infer the human relationships. In particular, the agent calculates human network, and measures the distance between the instances of the observer and the watched person by the ontology.

Next, Situation Recognizer Ag recognizes situations in detail, by acquiring information on the user's surrounding environment and applying background knowledge such as common sense. This knowledge is accumulated beforehand as ontology to recognize the situation.

Finally, Advisor Ag judges the providing service quality and the privacy level by considering the human relationship information from Relation Recognizer Ag and the user situation information from Situation Recognizer Ag. Then Advisor Ag directs Manager Ag. Manager agent decides the actual providing QoS to the user by cooperating with Device-aware agents.

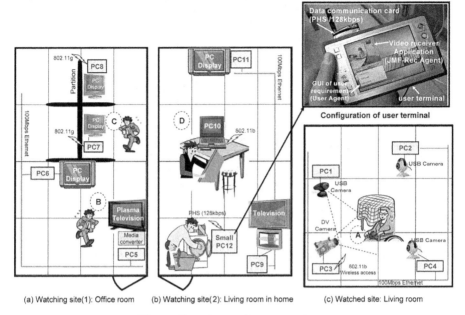

(a) Watching site(1): Office room   (b) Watching site(2): Living room in home   (c) Watched site: Living room

Fig. 3. Experimental room settings

4   Application: A Watching System for Elderly People

4.1   Implementation and Experimental Environment

We are developing an application of uEyes to watch over elderly people in home. We implemented agents based on AMUSE framework using DASH that is an agent-based programming environment based on ADIPS [10].

We introduce application scenarios to evaluate feasibility and effectiveness of our system. We suppose a situation where an elderly person is watched over by his son in remote sites. We have three experimental rooms as shown in Fig.3. One is supposed to be a living room in the watched person's home. This is a watched site. Others are regarded as his son's office and his living room in his home. They are watching sites. Fig.3(c) shows a room setting of the watched site. In terms of a location sensor, an active-type RFID system [14] is employed in this room.

Fig.3(a) and Fig.3(b) show the room settings of the watching sites. The son moves around the office room (a) and his living room (b). Here, "Small PC12" represents the watcher's user terminal which is shown in the right upper picture in Fig.3. This is a special device that is always brought with the observer. When other displays cannot be available, this terminal is selected for receiving the video of the watched person.

Also an User agent resides in this terminal to monitor the user's requirements and presence. In terms of the location sensor, ZPS ultra-sonic sensor [13] is used in both of the (a) and (b).

In this situation, our system selects the most appropriate camera, the PC with reasonable network connection, and the display devices, considering the multiple contexts. Then a live video with suitable quality is displayed on one of the displays according to his son's requirement on the watching over and the status of devices.

5   Experiments

5.1   Exp.(1): Experiments with Device-Aware and User-Aware Agents

In this experiment, firstly, a watching person specifies a user requirement from "best resolution" or "best smoothness" options, using a user interface on the user's terminal provided by the User Ag. This selection is based on the background of followings:

(1) The son wants to watch the father's facial color or expression in high resolution of the video, because he is worrying about the status of his father's sickness.

(2) The son wants to see the full-length of his father as smooth as possible, because he cares for his father's health condition.

To make the effectiveness of our system clear, we compare our system with a scheme in previous works, that is, location-based service configuration. This scheme selects the nearest camera and display to the watched/watching people, respectively, without any consideration of total quality of the service. This single-context-based configuration is actually tested by our system with terminating some agents that provide multiple contexts.

The son moves in the rooms. In this experiment, location of the father is fixed for simplification at point "A" in Fig.3(c). Based on the requirements and the location of the son, agents cooperatively work together to select the most adequate sets of entities. In this case, the son specifies "best resolution" of the video to watch his father's facial color in this experiment. Then he moves to the location at point "B" in Fig.3(a). The point "B" is the service area of a PC display of PC6 and a TV connected to PC5.

Here, in case of the location-based scheme, the video service migrated to the PC display of PC6 from the user terminal, because it is judged as the most nearest PC display, as shown in Fig.4(a). However, the quality of the video is too low to see the father's facial color vividly, because it moved with just the same video quality parameters as it was in the user terminal.

On the other hand, our system selected the TV display and the DV camera with high resolution to fulfill the user's requirement as shown in Fig.4(b). In terms of the software entity, DVTS [15] is selected because it can provide the high quality of video. In this case, the multiple contexts were effectively coordinated, and the user requirement was satisfied.

Next, the son specifies the high smoothness of movement of the video to watch in his father's health condition. Then he moves to the location at point "D" in

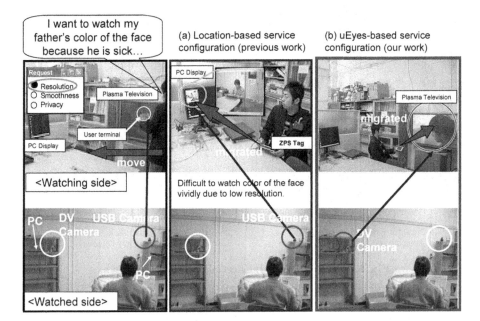

Fig. 4. Result of Exp.(1): Effect of Device-aware and User-aware agents in case of high quality requirement at point "B" in Fig.3(a)

Fig.3(b). The point "D" is the service area of a portable PC10 and a PC display of PC11.

Our system selected the PC display of PC11 and the USB camera connected to PC1, with high frame rate to fulfill the user's requirement. In terms of the network context, PC11 is the best because it is connected by a wired link with 100 Mbps. Moreover, agents recognized that PC11 cannot play DVTS video because DVTS software was not installed in PC11. This is the reason why the USB camera with the JMF-send Ag is selected. In this case, the multiple contexts were deeply considered, and the user requirement for high smoothness of the video was satisfied.

5.2   Exp.(2): Experiments with Social-Aware Agents

Next, we experimented to verify the effect when Social-aware agents are introduced. It is expected that it achieves the watching by a community of two or more people, because understanding of the interpersonal relationship and recognition of the situation can be done by the Social-aware agents.

Here, we show an example of watching by (a) the watched person's son, (b) his relative, (c) his neighbor with a good relation, and (d) a person living in the same block as him. In the situation when the watched person's privacy should be considered (Exp.(2)-1), for example when he is in the bedroom, each display for the person (a)-(d) is shown in Fig.5(A).

Also Fig.5(B) shows the display in the emergency situation when the watched person collapses onto the floor (Exp.(2)-2). Situation Recognizer judges it to be

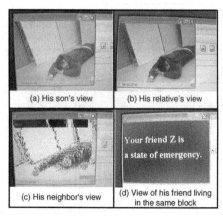

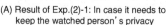

(A) Result of Exp.(2)-1: In case it needs to keep the watched person's privacy

(B) Result of Exp.(2)-2: In case of the watched person in emergency

Fig. 5. Result of Exp.(2): Effect of Social-aware agents

an emergency of the fall down etc., using the elderly person's location information by an ultrasonic tag, and background knowledge concerning the structure of the house or his daily lifestyle. For example in the case of Exp.(2)-2, the agent autonomously recognizes that he is in emergency, because the tag is staying in 10 cm high for over five minutes in the dining room. The display is changed so that many people may know the situation by lowering the privacy level. Concretely, an untouched image is delivered to the son and the relative, the neighbor with a good relation receives a low quality image that tells only the appearance, and the person of the same block is notified by the emergency message without any video image.

5.3   Evaluation

Effect of Device and User awareness: From the experiments described in the previous sections, we confirmed the feasibility of our proposal. We evaluated that our system can effectively construct service configuration that matches the requirement, coping with not only the location information, but also the device status around the users in the pervasive environment. In this application, heterogeneous entities like display devices, capture devices, PCs, networks, different kinds of sensors, software components, etc., are efficiently integrated.

Effect of Social awareness: From the experiments, we confirmed the effect of social-aware agents. By employing the agents, uEyes can recognize the relationship between people in real world and the detailed situation of the watched person. Based on the information uEyes can handle the QoS parameter and the privacy in adequate level. This feature would enable secure and safety watching-over services for non-expert users.

Effect of AMUSE: The integration of many entities was successful because of the introduction of an layered software architecture with the entities and agents.

This architecture is employed to accelerate the reuse of the framework and the middleware by other types of applications. Basically, the modularity, the autonomy and the loose coupling characteristics of the agents would meet to the construction of pervasive systems. It can adapt to diversity of types of entities and scalability of system size. By using this architecture, development and extension of the system will be easily accomplished.

6 Conclusion

In this paper we proposed a gentle system for watching over for pervasive care-support services. In this system, the situations of users and environmental information around him/her are effectively handled to provide the watching services. Social awareness such as human relationship is also considered to increase sense of safety in the watching.

In the future work, we will continue the case study in many different situations using this prototype system. Also we are trying to measure the user's satisfaction in quantitative manner.

Acknowledgement. This work was partially supported by the Ministry of Education, Culture, Sports, Science and Technology, Grants-in-Aid for Scientific Research, 17500029.

References

1. Yamazaki, T.: Ubiquitous Home: Real-life Testbed for Home Context-Aware Service. In: First International Conference on Testbeds and Reserch Infrastructures for the DEvelopment of NeTworks and COMmunities (TRIDENTCOM'05), pp. 54–59 (2005)
2. Minoh, M., Yamazaki, T.: Daily Life Support Experiment at Ubiquitous Computing Home. In: The 11th Information Processing and Management of Uncertainty in Knowledge-Based Systems International Conference (IPMU2006) (2006)
3. Camarinha-Matos, L.M., Afsarmanesh, H.: Infrastructures for collaborative networks - An application in elderly care. In: The 2005 Symposium on Applications and the Internet (SAINT'05), pp. 94–101(2005)
4. Sedky, M.H., Moniri, M., Chibelushi, C.C.: Classification of smart video surveillance systems for commercial applications. In: IEEE International Conference on Advanced Video and Signal based Surveillance (AVSS2005), pp. 638–643 (2005)
5. Consolvo, S., Roessler, P., Shelton, B.E.: The CareNet Display: Lessons Learned from an In Home Evaluation of an Ambient Display. In: The Sixth International Conference on Ubiquitous Computing (UbiComp2004), pp. 1–17 (2004)
6. Takemoto, M., et al.: Service-composition Method and Its Implementation in Service-provision Architecture for Ubiquitous Computing Environments. IPSJ Journal 46(2), 418–433 (2005)
7. Silva, G.C.D., Oh, B., Yamasaki, T., Aizawa, K.: Experience Retrieval in a Ubiquitous Home. In: The 2nd ACM Workshop on Capture, Archival and Retrieval of Personal Experiences (CARPE2005), pp. 35–44 (2005)

8. Cui, Y., Nahrstedt, K., Xu, D.: Seamless User-level Handoff in Ubiquitous Multimedia Service Delivery. Multimedia Tools and Applications Journal, Special Issue on MobileMultimedia and Communications and m-Commerce 22, 137–170 (2004)

9. Lohse, M., Repplinger, M., Slusallek, P.: Dynamic Media Routing in Multi-User Home Entertainment Systems. In: The Eleventh International Conference on Distributed Multimedia Systems(DMS'2005) (2005)

10. Fujita, S., Hara, H., Sugawara, K., Kinoshita, T., Shiratori, N.: Agent-based design model of adaptive distributed systems. Applied Intelligence 9, 57–70 (1998)

11. Takahashi, H., et al.: Design and Implementation of An Agent-based middleware for Context-aware Ubiquitous Services. Frontiers in Artificial Intelligence and Applications, New Trends in Software Methodologies, Tools and Techniques, vol. 129, pp. 330–350 (2005) In: The 4th International Conference on Software Methodologies, Tools and Techniques(SoMeT 2005), pp. 330–350 (2005)

12. Takahashi, H., et al.: AMUSE: An Agent-based Middleware for Context-aware Ubiquitous Services. In: The International Conference on Parallel and Distributed Systems (ICPADS2005), pp. 743–749 (2005)

13. http://www.furukawakk.jp/products/

14. http://jp.fujitsu.com/group/fst/services/ubiquitous/rfid/index.html

15. Ogawa, A., et al.: Design and Implementation of DV based video over RTP. In: Packet Video Workshop 2000 (2000)

A Context-Driven Programming Model for Pervasive Spaces

Hen-I Yang, Jeffrey King, Abdelsalam (Sumi) Helal, and Erwin Jansen

CSE Buliding Room E301, University of Florida, P.O. Box 116120,
Gainesville, FL 32611-6120, USA
{hyang,jck,helal,ejansen}@cise.ufl.edu
http://www.icta.ufl.edu

Abstract. This paper defines a new, context-driven programming model for pervasive spaces. Existing models are prone to conflict, as it is hard to predict the outcome of interleaved actions from different services, or even to detect that a particular device is receiving conflicting instructions. Nor is there an easy way to identify unsafe contexts and the emergency remedy actions, or for programmers and users to grasp the complete status of the space. The programming model proposed here resolves these problems by improving coordination by explicitly defining the behaviors via context, and providing enhanced safety guarantees as well as a real-time, at-a-glance snapshot of the space's status. We present this model by first revisiting the definitions of the three basic entities (sensors, actuators and users) and then deriving at the definition of the operational semantics of a pervasive space and its context. A scenario is provided to demonstrate both how programmers use this model as well as the advantages of the model over other approaches.

1 Introduction

As the field of pervasive computing matures, we are seeing a great increase in the number and the variety of services running in a pervasive space. As other researchers have found, a programming model for pervasive spaces is required, so that these environments can be assembled as software. The keystone of such a model is context, because of the dynamic and heterogeneous nature of these environments. Several systems and tools have been created to support context-aware development of pervasive spaces [1, 2, 3, 4].

While these existing programming models provide a convenient means for specifying the rules and behaviors of the applications with a formulated and systematic process, they do not provide an easy way to grasp the overall state of the smart space with a glance. In these models, obtaining the "big picture" of the space requires intentional effort; programmers first must know exactly what to look for, then explicitly query the system for particular data, and then compose and interpret the collected data, weaving them into an overall understanding.

As the number of applications grows, or the scenarios become more complicated, or services are provided by different vendors, conflict is almost certain. The existing models provide little support to detect conflicts. In these service-oriented models,

T. Okadome, T. Yamazaki, and M. Mokhtari (Eds.): ICOST 2007, LNCS 4541, pp. 31–43, 2007.

each service is focused on its own goals, ignoring the potential side-effects of concurrent access to shared resources – side-effects that could put the space into a dangerous or anomalous state.

These existing models also lack a simple way to identify, at a global level, unsafe contexts that should be avoided and the means to exit them if they arise. Instead, each service must have its own routines for detecting and handling these situations. This place an unrealistic demand on services developers, and safety guarantees in the space are infeasible, as a single missing or erroneous handler in a single service could allow the space to be stuck in one of these impermissible contexts.

Since our work in smart spaces has focused on creating assistive environments for seniors or people with special needs [5], the safety and reliability aspects of the programming model are especially critical. During the creation of our two smart homes – a 500 sq. ft. in-lab prototype, and the full-scale, 2500 sq. ft. house [6] – we found that the existing models and tools could not provide us with satisfactory safety guarantees and conflict management capability. We began expanding the ideas of context-oriented design to create a new programming model for pervasive spaces.

The main idea of our approach is to employ standard ontology to build a context graph that represents all possible states of interest in a smart space. Contexts in the graph are marked as desirable, transitional, or impermissible, and the goal is to take actions to always lead the smart space to desirable contexts, and to avoid impermissible contexts. At run time, the torrent of raw readings from sensors are interpreted and classified into active contexts, and the associated actions are set in motion to drive towards and strive to maintain desirable contexts.

Since services provided in a smart space using our model are programmed as various actions to be taken based on active contexts, it is easy to prevent and to verify no conflicting actions are associated with each context. Decisions on action plans based on the active contexts also avoid the danger that can be caused by executing interleaved actions from different services. The explicit designation of impermissible contexts allows the system to take emergency actions and exit the dangerous situations. These all contribute to the enhanced safety guarantees our model offers.

Finally, the explicit context graph and the visualization of currently active contexts allow real-time, at-a-glance snapshots of the space's status, which give the users and programmers alike an understanding of the overall picture of the smart space.

2   Related Work

The majority of ubiquitous computing research involving implemented systems has been pilot projects to demonstrate that pervasive computing is useable [7]. In general, these pilot projects represent ad-hoc, specialized solutions that are not easy to replicate. However, one thing that these applications do share is the notion of context.

Context-aware computing is a paradigm in which applications can discover and take advantage of contextual information. This could be temperature, location of the user, activity of the user, etc. Context is any information that can be used to characterize the situation of an entity. An entity is a person, place, or object that is considered relevant to the interaction between a user and an application, including the user and application themselves [8].

In order to ease the development of pervasive applications, effort has been placed into developing solutions that enable easy use of context. There are two main approaches: libraries and infrastructure. A library is a generalized set of related algorithms whereas an infrastructure is a well-established, pervasive, reliable, and publicly accessible set of technologies that act as a foundation for other systems. For a comparison between the two, see [9].

The Context Toolkit [1, 10] provides a set of Java objects that address the distinction between context and user input. The context consists of three abstractions: widgets, aggregators and interpreters. Context widgets encapsulate information about a single piece of context, aggregators combine a set of widgets together to provide higher-level widgets, and interpreters interpret both of these.

In the Gaia project, an application model known as MPACC is proposed for programming pervasive space [11], and includes five components with distinct critical functionalities in pervasive computing applications. The model provides the specification of operation interfaces, the presentation dictates the output and presentations, the adaptor converts between data formats, the controller specifies the rules and application logic and the coordinator manages the configurations.

EQUIP Component Toolkit (ECT) [12], part of the Equator project, takes an approach more similar to distributed databases. Representing entities in the smart space with components annotated with name-value pair property and connection, ECT provides a convenient way to specify conditions and actions of the applications, as well as strong support in authoring tools, such as graphic editors, capability browsers and scripting capabilities.

The use of ontology in our model is related to the notion of Semantic Web services [13]. The idea is that by giving a description of a web service we can automate tasks such as discovery, invocation, composition and interoperation. The Semantic Web makes use of description logic to describe a service. Description logics are knowledge representation languages tailored for expressing knowledge about concepts and concept hierarchies. An agent can use this description to reason about the behavior or effect of the invocation of a web service.

The SOCAM architecture [2] is a middleware layer that makes use of ontology and predicates. There is an ontology describing the domain of interest. By making use of rule-based reasoning we can then create a set of rules to infer the status of an entity of interest. The SOCAM research shows that ontology can be used as the basis of reasoning engines, but such engines are not suitable for smaller devices due to the computational power required.

CoBrA [3] is a broker-centric agent architecture. At the core of the architecture is a context broker that builds and updates a shared context model that is made available to appropriate agents and services.

Our programming model differs from the models and tools summarized in this section in that we give a formalization of the pervasive space and solely make use of description logic to interpret the current state of the world. This guarantees that contexts declared can actually occur in the space and prevents us from simultaneously activating contradictory contexts. We also explicitly describe the effect of entities in a smart space, allowing the system to detect conflicting or dangerous behaviors at compile time. This provides safety assurance in the pervasive space at all times.

3 The Physical World

Unlike traditional computing systems, which primarily manipulate a virtual environment, smart spaces deal with the physical world. We observe and interact with the world. We consider the world to be in a certain state at a given time. Smart spaces encode the physical world using three entities: sensors, actuators, and users.

Sensors and actuators are active objects in the space. Sensors provide information about a particular domain, giving the system information about the current state of the space. A sensor cannot change the state of the space – it can only observe. Actuators are devices that influence the state of the space. They can influence the state because the invocation of an actuator has at least one intentional effect on a particular domain. For example, an intentional effect of the air-conditioner actuator is to cool the room, affecting the "temperature" domain. This change can be observed by a temperature-domain sensor, such as a thermometer.

Objects other than sensors and actuators are "passive" or "dumb" objects, and cannot be queried or controlled by the space. These entities, such as sofas or desks, are therefore irrelevant to the programming model. They may be embedded with active objects (e.g., a pressure sensor in the seat of a couch), but it is the sensors or actuators that are important, not the passive object itself.

Users are special entities in the space. While they are of course not "wired" into the system like sensors or actuators, users are an integral part of the smart space, as it is their various and changeable preferences and desires that drive the operation of the space.

3.1 Observing the World with Sensors

We consider our world of interest to be $U = \prod D_j$ where D_j is a domain, within the bounds of the space, that is observable by sensors. We will use $u \in U$ to denote aspects of the current state of the world.

Sensors provide data about the current state of affairs. A smart space typically contains many physical sensors that produce a constant stream of output in various domains. As this data is consumed by the smart space, each physical sensor can be treated as a function the space can call to obtain a value.

Definition 1 (Sensor). A sensor is a function that produces an output at a given time in a particular domain.

$$f_i^{\,j} : U \rightarrow D_i$$

Notice that we can have multiple sensors observing a single domain D_i. This is indicated by $f_i^{\,1}, f_i^{\,2}$, etc. Sensors in the same domain may produce different values due to different locations, malfunction or calibration issues. The smart space system will be responsible for correctly interpreting data. We can group all available sensors together and define $f = \prod f_i$ to be a snapshot of all sensors at a point in time.

3.2 Influencing the World with Actuators

Actuators allow the smart space to affect the physical world. We model the actuators as a set A with elements $a_i \in A$. Our model assumes that while an actuator is turned off it has no effect on the world. In this sense, an actuator that is off is the same as an actuator that does not exist in the space.

Every actuator in the space has certain intentional effects, which are the goals we intend to achieve with activation of the actuator. Programmers specify the intentional effect by defining how activation of the actuator will affect the state of the pervasive space – in other words, given the current state, to what new state will the space transition when the actuator is turned on. We formalize the intentional effect of an actuator as follows:

Definition 2 (Intentional Effect). An intentional effect of an actuator a is the set of states that possibly can arise due to invocation of that actuator:

$$g_a : U \rightarrow 2^U$$

An actuator may have more than one intentional effect. For example, turning on a vent can be used both to improve the air quality and raise the overall power draw.

3.3 Inhabiting the World as a User

To model the user we roughly follow the belief-desire-intention [14] model, in the sense that the user observes the state of the space and, based upon this information, commits to a plan of action. A user's *Desires* are the set of preferences about the state of the world. These preferences dictate how the world should be at that given moment. *Belief* is the current observed state of the world, interpreted by taxonomy. Ideally the system's perception of the state of the world should the user's perception. *Intention* is the action plan to which the user commits. In a smart space this is the activation of a set of actuators, aiming to fulfill the desire as previously mentioned.

The idea is that the user observes that state of the world. The user then classifies the state according to the users' specific taxonomy. This classification gives rise to a set of contexts that are active, and each context is associated with a set of behaviors. The intention of these behaviors is to change the current state to a more desirable state. For example, if user interprets the world as *Cold Indoors,* we turn on the heater with the intention to reach the desired context of *Warm Indoors.*

4 The Context-Driven Programming Model for Pervasive Spaces

Our smart space programming model involves interpreting the sensor readings, controlling the actuators and capturing and reasoning about the users' desires.

4.1 Interpreting Sensor Data

Dealing directly with sensor data is rather awkward. Instead, we would like to work with higher level information that describes the state of the world. A natural choice for describing knowledge is to use description logics. Using description logic we can define an ontology that describes the smart space. We will restrict ourselves to taxonomic structures, and therefore will have no need to define roles. The language L we define here is similar to ALC without roles. This language, and description logics in general, are described in more detail in [15]. We will closely follow their definitions, but will use the term context and concept interchangeably:

Definition 3 (Taxonomy). Concept descriptions in L are formed using the following syntax rule:

$$C, D \rightarrow AC \mid \top \mid \bot \mid \neg C \mid C \sqcap D \mid C \sqcup D$$

where AC denotes an atomic concept.

Atomic concepts are those concepts that are directly associated with sensor readings, and are not defined in terms of other concepts. For instance, we can define the atomic concepts *Smokey*, *Clear*, *Smelly* and *Fresh*. The first two concepts can be directly observed by a smoke detector and the last two can be observed by chemical sensor. We could then construct derived concepts, such as *Murky_Air = Smokey* $\sqcap$ *Smelly*.

Apart from describing concepts we need to have a way to interpret these concepts, as this will allow us to interpret the state of the space $u \in U$. Hence we define concepts in terms of U:

Definition 4 (Interpretation Function). *We define the interpretation I as follows:*

$$AC^{I} \subseteq U \qquad\qquad \top^{I} = U \qquad\qquad \bot^{I} = \varnothing$$
$$(\neg C)^{I} = U - C^{I} \qquad (C \sqcap D)^{I} = C^{I} \cap D^{I} \qquad (C \sqcup D)^{I} = C^{I} \cup D^{I}$$

We would also like to specify how a context or group of contexts relates to another. There are two ways of declaring relationships between two contexts. One is the inclusion relation, where some base context is a component of a larger composite context (e.g., the base contexts *Smoky Air* and *Foggy* are included in the composite context *Smoggy*). The other is the equivalent relation, used to resolve ontological issues where the same context has multiple names. Formally these relations are defined as:

Definition 5 (Context Relations). Inclusion and equality are defined as follows:

Inclusion: $C \sqsubseteq D$, *interpreted by* $C^{I} \subseteq D^{I}$
Equality: $C \equiv D$, *interpreted by* $C^{I} = D^{I}$.

The description of how a set of concepts relate to each other is called a terminology or T–Box. A terminology T is a collection of inclusions and equalities that define how a set of concepts relate to each other. Each item in the collection is unique and acyclic. Concepts that are defined in terms of other concepts are called derived concepts. An interpretation I that interprets all the atomic concepts will allow us to interpret the entire taxonomy.

Interpretation of the current state of the world is straightforward. We identify whether the current state of the world is a member of any concept. In other words, to verify whether $u \in U$ satisfies concept C we check that $u^{I} \in C^{I}$. The interpretation of the current state of the space leads us to the current active context:

Definition 6 (Active Context). The active context of the space R: $U \rightarrow 2^{C}$ is:

$$R = \{C \mid u^{I} \in C^{I}\}$$

Much of the related research takes different approaches to derive higher level information. Most sensor networks in the literature make use of a hierarchy of interpretation functions, often referred to as context derivation [16, 17, 18], which take as input the readings (or history of readings) from sensors and produce a new output value. We avoid these derivations because there is no guarantee that the derived functions will follow the consistency rules as specified by description logic. The inconsistency implies that contradictory or invalid (undefined) context states could arise, introducing potentially dangerous situations.

4.2 Controlling Actuators

Apart from observing the space, we are also controlling the space using actuators. Within a smart space, users and the system can perform sequences of actions by turning actuators on or off.

Definition 7 (Statement Sequence). Given a set of actuators A, a statement sequence S, also known as a sequence of actions, is defined as follows:

$$S ::= \uparrow a_i \mid \downarrow a_i \mid S1; S2$$

Where $\uparrow a_i$ turns actuator $a_i \in A$ on, and $\downarrow a_i$ turns actuator a_i off. We also denote action prefixing using the symbol ';' to indicate that we first perform statement S1 and after which we execute the remaining statement S2.

Using the description of an intentional effect it is now straightforward to identify whether or not two actuators conflict with each other. Two actuators a_i and a_j are in conflict in context u if $g_i(u) \cap g_j(u) = \varnothing$.

That is, the invocation of a_i leads to a state of the world that is disjoint from the state to which the invocation of a_j will lead, hence this state of the world can never arise. A classical example of the invocation of two conflicting actuators is the air-conditioning and the heater. A heater will raise the room temperature whereas an air-conditioner will lower the temperature, hence a statement that activates both the air-conditioner and the heater at the same time is considered to be contradictory.

4.3 Beliefs-Desires-Intentions of Users

As we mentioned earlier a user has a belief about the possible states of the world T, which can be captured by taxonomy and the interpretation of the user. The current belief about the state of the world can be captured with taxonomy and an accompanied interpretation. The interpretation maps the values obtained from a sensor to the atomic concepts. We express the intentions I of the users by a sequence of actions the user wishes to execute in a particular context. Formally, a user can be denoted by

Definition 8 (User). A user can be denoted by a tuple

$$B ::= <T ; I; D, X, I>$$

Where T and I are the taxonomy accompanied by the interpretation of the user. The user can then specify desired contexts D, and extremely undesirable impermissible contexts X. Let $I: C \to S$ be a mapping from context to statements that shows possible intentions to transit away from context C.

Definition 9 (Active and Passive Intentions). Assume current active context is represented by R the following transitions defined the activeness and passiveness of intentions:

$$\textbf{Active Intention:} \quad I_a : R \xrightarrow{I_a} D$$

$$\textbf{Passive Intention:} \quad I_p : D \xrightarrow{I_p} R$$

For instance, turning on a ventilation fan is an action that has an intentional effect of transiting the air quality in a house from *Murky Air* to *Clean Air*, therefore "Turn On Vent" is an active intention. Turning on the stove to sear steak, on the other hand, can produce smoke and degrade air quality from *Clean Air* to *Murky Air*. By turning off the stove we may eliminate the source of the smoke, therefore there is a possibility for improving the air quality over time, hence "Turn Off Stove" is a passive intention. In our model, the active intentions (what happens when an actuator is turned on) are specified by programmers, and the passive intentions (what happens when an actuator is turned off) are inferred by the system.

4.4 The Smart Space Programming Model

We can now define a space consisting of sensors and actuators and a user using the following operational semantics:

Definition 10 (Programmable Pervasive Space). A programmable pervasive space can be denoted by a tuple consisting of:

$$P ::= <B;R;\ S;\ 2^A>$$

Where B is the representation of the user, R the currently active context, S the statements we are currently processing and 2^A the set of actuators currently active.

Spaces changes over time due to nature. We model the effect of nature by a transition that changes the state of the space, and observe the changes of the space through our sensors.

Definition 11 (Environmental Effect)

$$<b;\ f(u);\ S;\ a> \ \rightarrow\ <b;\ f(u');\ S;\ a> \text{ where } f(u) \neq f(u')$$

The other way in which the space can change is by actuators that the space controls. The following transitions capture the turning on and off of actuators:

Activation: $<b;\ u;\ \uparrow ai;\ a> \ \rightarrow\ <b;\ u;\ \varepsilon\ ;\ a \cup ai>$
Deactivation: $<b;\ u;\ \downarrow ai;\ a> \ \rightarrow\ <b;\ u;\ \varepsilon;\ a \setminus ai>$

If the set of active contexts has changed, we obtain the intentions from the user and execute the desired sequence of actions:

Context Change: $<b;\ f(u);\ S;\ a> \ \rightarrow\ <b;\ f(u');\ i(R(f(u')));\ a>$ whenever $R(f(u')) \neq R(f(u))$.

5   Scenario and Programming Procedures

The context-driven model defined in the previous sections, when comparing to other existing programming models for pervasive computing, provide stronger safety features and better capability to evaluate multiple potential action plans. The overall state of a smart space is captured in a context graph based on the interpretation of the user (T, I), in which the desirable contexts (D) and the impermissible contexts (X) are specified. The sensors retrieve the readings, and identify the current active contexts (R), and then the smart space devises the plan I to move the state from R toward D. We next demonstrate three advantages of this model using the following scenario.

5.1   Applying the Context-Driven Programming Model

Matilda is an 85-year old lady, living alone in a free-standing smart home. Though suffering from mild dementia, she does not require live-in assistance. Instead, she makes use of the services in the smart house that provides cognitive assistance. On a chilly night in December, Matilda is searing a rib eye steak on the stovetop for dinner. Her steak is well-marbled with fat, and is generating a lot of smoke as it cooks. The smoke clouds up the kitchen, and soon sensors detect that the air quality in the house is fast degrading. Matilda is unaware that there is a problem, but the services in the smart house can alert her to the dangers, cue her for response, or even resolve the problems automatically. The remainder of this section describes the development of these cognitive assistance services using our context-driven programming model.

According to the context diagram of the house (Fig. 1), there are two contexts in the "Air Quality" domain ($D_{air\_quality}$). They are the *Murky Air* context and the *Clean Air* context, with *Clean Air* being the preferred context. The air quality can be monitored using a Smoke Detector Sensor, represented in our model as the function:

$$f_{smoke\_detector} : U \rightarrow D_{air\_quality}$$ (*U* represents the overall state of the house)

The current state of the house can be represented by the active context of the space R, which consists of the contexts that can be observed and are currently true. Before Matilda starts to sear the steak, R = {*Clean Air, Low Power Draw, Warm Indoors*}. At some point, because of Matilda's cooking, *Murky Air* has replaced *Clean Air* in *R*.

Looking up the context graph, as shown in Fig.1, the home server found three possible courses of actions that can be employed to improve the quality of the air: opening the window, turning on the vent, or turning off the stove.

For this particular example, there are three actuators of interest. Therefore we define the actuator set A as {*Window, Vent, Stove*}, because at least one of their intentional effects are relevant to air quality. In particular, "Open Window" and "Turn On Vent" can both take us actively from the *Murky Air* to *Clean Air* context, while "Turn On Stove" has the reverse effect, therefore it is possible that by turning *off* the stove, we can observe the improvement of air quality. These intentional effects are described in our model as: $g_{window}^{air\_quality}$: *Murky air* → *Clean air*, $g_{vent}^{air\_quality}$: *Murky air* → *Clean air*, and $g_{stove}^{air\_quality}$: *Clean air* → *Murky air*. In this case, "Open Window" and "Turn on Vent" are active potential actions because they can proactively improve the air quality, while "Turn Off Stove" is a passive option.

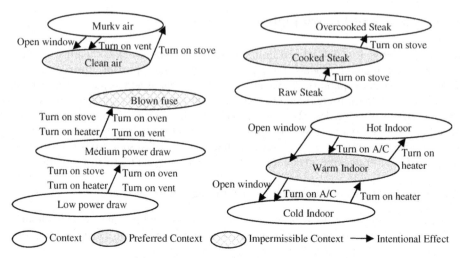

Fig. 1. Context Graph for Matilda's House

5.1.1 Evaluating Potential Actions

When evaluating all the possible options, the home server first examines the active options before considering passive options. It starts by checking the side effects of each of the potential actions. In this case, it finds that, in addition to improving the air quality, opening the windows may cause the room temperature to drop if the outdoors temperature is lower than the current room temperature. It also finds that turning on the vent will increase the power load of the house. These side effects are as:

$$g_{window}^{temperature} : Hot\ Indoors \rightarrow Warm\ Indoors;\ Warm\ Indoors \rightarrow Cold\ Indoors$$

$$g_{vent}^{power\_draw} : Low\ Draw \rightarrow Medium\ Draw;\ Medium\ Draw \rightarrow Blown\ Fuse$$

Upon calculation, the home server decides that opening the window would cause the room temperature to drop from warm to cold, which is much less preferable, while turning on the vent will only increase the power draw marginally. Hence the home server decides that turning on the vent is the preferable action. In other words, the home server will now choose an action plan consisting of the statement $S = \{\uparrow vent\}$ in an attempt to improve air quality in the house to *Clean Air*.

5.1.2 Detecting Conflicting Directives

A few minutes pass, but the sensors have not reported any significant improvement in air quality. The vent can barely keep up with the smoke produced by searing the steak. The context *Murky Air* is still part of the active context R, and the home server must employ another means to clear the air. Its only remaining option is to turn off the stove ($S_1=\{\downarrow stove\}$) so as to stop the steak from producing more smoke. However, this is in conflict with the cooking assistance service, which wants to keep the stove on high heat ($S_2=\{\uparrow stove\}$) until the internal temperature of the steak reaches 133 °F.

As we have defined, a pervasive space is represented as the 4-tuple $\{B;\ R;\ S;\ 2^A\}$. At this moment, S includes two statements: S_1 (trying to deactivate the stove actuator) and S_2 (trying to activate the same stove actuator). Detecting that S_1 and S_2 are contradictory, the home server prompts Matilda to see if she prefers to turn off the stove to improve the air quality or to leave the stove on until the steak is done.

5.1.3 Avoidance and Handling of Impermissible Contexts

Matilda decides that just steak isn't enough for dinner, and prepares to toast some garlic bread in the oven. She would of course like to toast the bread while the steak is searing, so everything is done at the same time.

On this chilly night, however, the heater in Matilda's house is already running at full force. With the stove already turned on to high heat and the vent whirling to keep the smoke out of the kitchen, the home server calculates that turning on the oven would draw too much power, resulting in an impermissible context, *Blown Fuse*. In such a cold night, Matilda's frail condition, a blown fuse that cut off electricity would greatly endanger her health. The home server therefore decides to prevent oven from being turned on, and announce a message to Matilda through the speaker in the kitchen to try again later.

5.2 Programming Procedures

How do programmers actually program a pervasive computing space using context-driven programming model? We identified the following three-step procedure:

1. **Design the Context Graph:** Programmers have to decide what the domains of interest are, and what the contexts of interest within these domains are. This decision is heavily influenced by availability of the sensors, the services planned, and the users' belief and desires.
2. **Interpret Sensor Readings:** Programmers have to define interpretation functions from ranges or enumerated possible reading values from various sensors to atomic contexts appearing in the context graph.
3. **Describe Intended Behaviors:** Programmers have to describe intended behaviors in terms of action statements associated with each context in the context graph, so that smart space knows which actuators to manipulate when various contexts become active.

A prototype IDE has been implemented to facilitate the programming practice using this context-driven model. The prototype is currently being tested internally.

6 Conclusion

Many research projects have explored different programming models for pervasive computing, but the applications developed can only focus on achieving their own goals and often lose sight of the overall picture in the smart space. This can lead to conflicts and erroneous behaviors. Building upon the experience learned during the implementation of the Gator Tech Smart House, we proposed and experimented with a new context-driven programming model to address some of these shortcomings.

In other existing models, contexts only complement the traditional programming; our model uses contexts as the primary building blocks. Programmers build the context graph that captures all possible states that are of interest in the smart space.

Contexts from the graph are marked as desirable, transitional, or impermissible. Programmers also define the intentional effects of actuators in terms of transitions from one context to another. The system is responsible for identifying active contexts from sensor readings and choosing actions that can lead to more desirable contexts.

The benefits of our context-driven programming model are improved coordination using explicitly defined behaviors based on context, enhanced safety guarantees and real-time, at-a-glance snapshots of the space's status. By explicitly describing the effect of all the basic entities we can detect conflicting devices at compile time, and the system is better able to evaluate multiple potential action plans. In addition, this explicit description makes it easier to change services based on users' varying preferences. The model is also able to detect conflicting or dangerous behaviors at runtime. Finally, the explicit context graph allows programmers to define impermissible contexts – states of the space that are extremely dangerous and must be avoided – as well as providing the support at runtime to avoid and handle them.

The formalization of the model and the provided scenario demonstrate the practicality of this model. We are currently developing and testing tooling support for this model, as well as integrating it with other technologies developed in our lab, mainly the Atlas sensor network platform [19], providing an end-to-end solution for creating programmable pervasive spaces.

References

1. Dey, A., Salber, D., Abowd, G.: A conceptual framework and a toolkit for supporting the rapid prototyping of context-aware applications. Human-Computer Interaction (HCI) Journal 16, 97–166 (2001)
2. Gu, T., Pung, H., Zhang, D.: A service-oriented middleware for building context-aware services. Journal of Network and Computer Applications (JNCA) 28, 1–18 (2005)
3. Chen, H., Finin, T., Joshi, A., Perich, F., Chakraborty, D., Kagal, L.: Intelligent agents meet the semantic web in smart spaces. IEEE Internet Computing, vol. 8 (2004)
4. Gu, T., Pung, H., Zhang, D.: Toward an OSGi-Based Infrastructure for Context-Aware Applications. In: IEEE Pervasive Computing, pp. 66–74 (October-December 2004)
5. Bose, R., King, J., El-zabadani, H., Pickles, S., Helal, A.: Building Plug-and-Play Smart Homes Using the Atlas Platform. In: Proceedings of the 4th International Conference on Smart Homes and Health Telematic (ICOST), Belfast, the Northern Islands (June 2006)
6. Helal, A., Mann, W., Elzabadani, H., King, J., Kaddourah, Y., Jansen, E.: Gator Tech Smart House: A Programmable Pervasive Space, IEEE Computer magazine, pp. 64–74 (March 2005)
7. Chen, G., Kotz, D.: A survey of context-aware mobile computing research. Technical Report TR2000-381, Dept. of Computer Science, Dartmouth College (2000)
8. Salber, D., Dey, A., Abowd, G.: The context toolkit: Aiding the development of context-enabled applications. CHI, pp. 434–441 (1999)
9. Román, M., Hess, C., Cerqueira, R., Ranganathan, A., Campbell, R., Nahrstedt, K., Gaia: A Middleware Infrastructure to Enable Active Spaces. In IEEE Pervasive Computing, pp. 74–83 (October–December 2002)
10. Greenhalgh, C., Izadi, S., Mathrick, J., Humble, J., Taylor, I.: ECT: A Toolkit to Support Rapid Construction of UbiComp Environments. In: Online Proceedings of the System Support for Ubiquitous Computing Workshop at the Sixth Annual Conference on Ubiquitous Computing (UbiComp 2004) (September 2004)

11. Wooldridge, M.: Reasoning about Rational Agents. The MIT Press, Cambridge, Massachussetts/London, England (2000)
12. Baader, F., Calvanese, D., McGuinness, D., Nardi, D., Patel-Schneider, P. (eds.): The Description Logic Handbook. Cambridge University Press, Cambridge (2002)
13. Chen, G., Kotz, D.: Solar: An open platform for context-aware mobile applications. An informal companion volume of short papers of the Proceedings of the First International Conference on Pervasive Computing, pp. 41–47 (2002)
14. Cohen, N., Lei, H., Castro, P., Purakayastha, A.: Composing pervasive data using iql. In: Proceedings of the Fourth IEEEWorkshop on Mobile Computing Systems and Applications, IEEE Computer Society, 94 (2002)
15. Kumar, R., Wolenetz, M., Agarwalla, B., Shin, J., Hutto, P., Paul, A., Ramachandran, U.: Dfuse: a framework for distributed data fusion. In: Proceedings of the first international conference on Embedded networked sensor systems, pp. 114–125. ACM Press, New York (2003)
16. King, J., Bose, R., Yang, H., Pickles, S., Helal, A.: Atlas: A Service-Oriented Sensor Platform, To appear in the Proceedings of the First International Workshop on Practical Issues in Building Sensor Network Applications (in conjunction with LCN 2006) (November 2006)

Tracking People in Indoor Environments

Candy Yiu and Suresh Singh

Portland State University
{candy,singh}@cs.pdx.edu

Abstract. Tracking the movement of people in indoor environments is useful for a variety of applications including elderly care, study of shopper behavior in shopping centers, security etc. There have been several previous efforts at solving this problem but with limited success. Our approach uses inexpensive pressure sensors, placed in a specific manner, that allows us to identify multiple people. Given this information, our algorithm can track multiple people across the floor even in the presence of large sensor error. The algorithm we develop is evaluated for a variety of different movement patterns that include turning and path crossing. The error in correct path detection is shown to be very small even in the most complex movement scenario. We note that our algorithm does not use any a priori information such as weight, rfid tags, knowledge of number of people, etc.

1 Introduction

The problem of tracking people in indoor spaces is an important component for a variety of applications including in-home care for disabled or seniors, etc. Traditional approaches to this problem include using multiple cameras, sound/vibration sensors, and RFID tags or other radio devices planted somewhere on the clothing. The problem with the latter solutions is that they lack generality while the camera solution tends to be expensive both in computation as well as cost. The solutions that have been implemented using sound/vibration require special raised floors for implementation and even with that requirement, they often display a significant lack of accuracy as well as inability to track multiple people. In this research we develop an algorithm for tracking multiple people simultaneously using simple pressure sensors that can be embedded in carpets or floor tiles. The algorithm can track people even when they are shoulder to shoulder and in the presence of arbitrary turns and path crossings.

The remainder of the paper is organized as follows. We discuss the sensor used as well as its sensing capability in section 2. We also characterize the error in locationing as measured for this sensor. In section 3 we examine the problem of sensor placement with the goal of low cost and high accuracy. Section 4 then presents our tracking algorithm. The tracking algorithm is studied in section 5. Related work is discussed in section 6 and conclusions are in section 7.

T. Okadome, T. Yamazaki, and M. Mokhtari (Eds.): ICOST 2007, LNCS 4541, pp. 44–53, 2007.
© Springer-Verlag Berlin Heidelberg 2007

2 Sensor Description

The sensors we use are inexpensive Flexiforce pressure sensors. They are as thin as a piece of paper (0.127mm) so that people cannot feel the existence of the sensor when they walk on the tile. The maximum sensing range is 454kg. Other specifications are [10]: length 203mm, width 14mm, sensing area 9.53mm, connector 3-pin Male. In use, we place these sensors directly underneath flooring tiles. The readings from multiple sensors (up to 8) are simultaneously fed into a standard serial port.

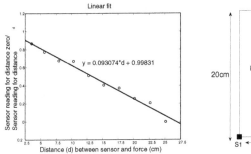
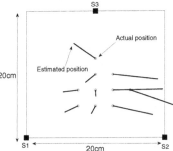

Fig. 1. Characterizing sensor readings   **Fig. 2.** Locationing error

To understand how the sensors react to force applied to the tile as well as the locationing error which comes about due to absorption by the tile (by flexing) and we characterize how sensor readings relate to the distance between the point on the tile where the force is applied and the location of the sensor underneath the tile. We place a sensor under a corner of a $20cm \times 20cm$ tile and applied force at various points on the tile. We obtain the sensor readings s_i for distance d_i. We also apply the same force on the tile directly on top of the sensor giving us reading s_0. Figure 1 plots s_0/s_i as a function of d_i. As we can see, there is a good linear fit between these quantities and the linear equation gives us the needed expression for interpreting the sensor readings.

We next place three sensors underneath the tile as shown in Figure 2. The goal here is to determine the position of an applied force without prior knowledge of the s_0 value. The unknowns therefore are s_0 and d^j ($j = 1, 2, 3$ for the three sensors). The measured values are s^j and we know the relative location of the three sensors. We can thus solve for the d^j values to determine where on the tile the force is applied. In the figure, we indicate by a 'o' the actual location of the force for each of the ten values shown and by a straight line the error between the actual position and the estimated position (the other endpoint of each line gives us the estimated position using sensor readings). The average error is 3.8cm though in some cases it is significantly more. Given that the tile is $20cm \times 20cm$, we can estimate the error as $\pi 1.5^2/64$ or approximately 12%.

3 Sensor Placement

Consider a high enough density of sensor deployment such that $m \geq 1$ sensors will be stepped upon by each foot. This gives us an accurate estimate of the foot's location but is very expensive. Let us consider an alternative technique which loses some accuracy but has a significantly reduced cost. We cover the floor with tiles of the size equal to an average foot step and place one sensor under each tile. Thus, when a foot is placed fully on or even partially on a tile, all the sensors under the tile sense some pressure. However, if there are two feet on the same tile, we have an ambiguous result as shown in Figure 3 since the data can be interpreted to mean that there are either three feet on three different tiles or two feet each of which overlaps two tiles. This ambiguity causes tracking error. Our goal is to be able to locate the foot step of each person while minimizing the total cost. There are a very large number of possible sensor/tile placements. However, these placements need to satisfy two requirements: first, the placements should ensure that at least one sensor senses each foot step otherwise we may fail to identify all foot steps of a person; and second, no more than one foot should step on each tile at any given time in order to ensure that we can distinguish between different foot steps.

    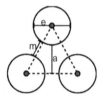

Fig. 3. One sensor per foot deployment ambiguity

Fig. 4. Circular tile with gap deployment

The set of possible sensor/tile placements can be divided into two – tiling that covers the floor completely or tiling that leaves gaps. We observed that the latter placements typically gave us far lower cost. In addition, these types of placements tend to reduce ambiguity where one foot can step on multiple tiles simultaneously. However, we need to be careful about how the tiles and sensors are placed so that we meet the two requirements mentioned previously.

Based on an evaluation of several placements, we came up with the deployment shown in Figure 4 with 3 sensors per tile. Assume that there is a minimum distance ϵ between two people. Thus, if ϵ is greater than the foot size, the possibility of two people stepping on three consecutive tiles as in Figure 3 will not happen. Let us assume that the foot step size is $l \times m$, where l is the length and m is the width of an average foot. We choose the diameter of the circular tile to equal ϵ because it can ensure that no two or more steps on the same tile belong to different people. Tiles are placed distance m apart which ensures that we will not miss any foot step. To ensure that at least one sensor senses every step, we require $a \leq l$ as shown in Figure 4. Let $m \approx 0.5l$ and $a = .5\epsilon - (\sin 60 \times (\epsilon + m))$.

Therefore, $\epsilon < 1.5l$. This means that ϵ has to be less than one and a half foot length, which is reasonable.

4   Tracking Algorithm: Sliding Window LMS

Assume that the locationing algorithms discussed previously give us foot locations for one or more people, with some error. The problem then is to track people as they move around the sensed area. We make no assumptions about how the people move except that physically impossible cases (such as one person walking through another) do not occur. We initially considered a simple algorithm which finds a straight line fit using a least mean square error metric. Unfortunately, in many cases, this algorithm has an unacceptably high error where it combines foot locations belonging to different people. Therefore, we enhanced the algorithm by using more information such as step length to distinguish between different people. It performs much better but it fails when we have crossing paths and turns. This led to the sliding window algorithm that explicitly finds turns in paths.

The algorithm is based on the assumption that a person's path consists of straight line paths and turns alternating with each other. Therefore, finding where the turns occur and fitting multiple straight lines to the provided foot locations gives us a good estimate of the paths. The algorithm is divided into two parts: an initialization stage and then an iterative stage.

The initialization stage considers the first three step locations of n people. Let p_1, p_2 and p_3 be the first three consecutive points and ζ be the default average step length. Let $e(p_i, p_j)$ be a function of the distance from p_i to p_j and $d(p_i, p_j, p_k)$ be a function of the shortest distance from point p_j to straight line p_i and p_k. The error of each three points is $d(p_1, p_2, p_3) + |\zeta - e(p_1, p_2)| + |\zeta - e(p_2, p_3)|$.

Consider all the possible three tuples consisting of the first, second, third steps. For each collection of n such tuples, we compute the sum of the least mean square error. Then the combination with the smallest error represents the first three steps of n people.

In the iterative stage, there are two important parts. The first part is to distinguish each point in the new set of points belonging to each path. The second part is to determine if the new point is a turning point for each path.

To achieve the first part, we extend the straight line formed from the initial stage by adding one more point (i.e., the fourth point). This is because people mostly walk in one direction (i.e., turns are infrequent). There are n points and n lines. For each person, the average step length is stored and updated in every iteration. It's initial value is ζ and it's value is re-calculated using $\frac{d(p_1, p_2) + d(p_2, p_3)}{2}$, which gives us an increasingly accurate estimate of step length as more data is collected. We take the combination of one line and one point and compute the error of fitting all the points to the line and the error of the average step length. Then, we calculate the sum of the mean square for all n people and we pick the combination that yields the lowest error.

In the second part of the iterative stage, we create a window per person (path) and only consider points that lie within a variable window. The window expands whenever a new point is added. However, it will slide and reduce in size when the tracking algorithm determines that the person has turned.

To understand the workings of this algorithm, it is useful to model a person's path as either being one where the person is walking in a straight line or the person is making a turn. The initial state for the algorithm starts with the assumption that the person is walking in a straight line. When in this state, the window expands by adding new points and adapting the paths. However, when a turn is detected, the window shrinks to 3 points only and then starts to grow again by adding additional points if the person is detected to be walking in a straight line otherwise it keeps sliding while maintaining a size 3.

Detecting turns works as follows. When people walk, their left and right feet hit the floor alternatively. This information is used by the algorithm to determine when a person turns. The key idea here is illustrated in Figure 5 where we see seven points corresponding to the path followed by one person (the points correspond to the left foot then the right then the left and so on). Let us look at points 1, 2, 3 first. Point 4 lies within the angle formed by the line segments connecting points 1-2 and 1-3. Therefore, point 4 does not represent a turn. Likewise, to determine whether point 5 represents a turn, we look at the angle formed between segments 2-3 and 2-4. Again, we see that point 5 lies well within this angular region. Consider point 6 now. We see that point 6 lies outside the angle formed by 3-4 and 4-5 and thus it represents a turning point. Until this point our window contained all the points from 1 to 5. However, after the turn is detected, we shrink it to points 4, 5, 6 only. Next we look at point 7. Since it lies within the angular region formed by segments 4-5 and 4-6, we conclude that the person is walking in a straight line again and the window grows to include points 4,5,6,7.

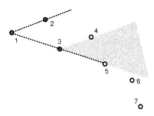

Fig. 5. Turning point

5   Simulation Results

In order to evaluate the tracking algorithm, we developed a test case generator which generates different topological test cases corresponding to varying numbers of people, location errors, etc. The output files contain a set of points in continuous time. In order to generate realistic test cases, we rely on a study of the walking behavior of people. [9] provides us the relationship between height and stride length $Height \times x = stridelength$ where $x \approx 0.4$. If we assume that the average height is 168cm, we get a stride length of 67cm.

5.1   Experimental Design

In order to evaluate delete our tracking algorithm, we designed an extensive set of experiments that gave us a total of 180 distinct test cases and each test case was repeated ten times. The variable parameters we use are

- Number of people: this number is varied from 2 to 5.
- Location error: this corresponds to error in determining the exact foot position and takes values 0%, 10%, 20%, 30% and 40%. Recall that we measured the locationing error to be approximately 12% (section 2). The erroneous location is generated as follows. The test case generator first generates the actual foot position. We then generate a uniform random angle between 0 and 2π and a random length equal to the error percentage times the average foot length. This position is the location fed to the tracking algorithm.
- Phase: people are either all in phase or out-of-phase (in-phase means they start with the same foot). For three or more people, each individual is out of phase with respect to the two on either side when we consider out-of-phase experiments.
- Direction: for two people, there are two cases (same direction or opposite direction). For three or more, we consider the cases when either all the people are going in the same direction or when the directions alternate.
- Path shape: We consider straight line with zero or one turn per person. For two people we consider all the cases (i.e., turns with the same or different direction). However, for three or more people, we consider only a subset of cases because more complicated cases can be reduced to a union of cases with two people.
- Crossings: We consider cases when straight line paths cross or turning paths cross. For two people we consider straight lines crossing at different crossing points as well as crossing with turns, again at different crossing points. We also consider a case when the two people turn but do not cross each other.

Tables 2,3 list some of the various test cases we study. In the case of two people, we consider all the possible combinations of turning, crossing, direction, and phase. In (a) the two individuals are in-phase but they are out-of-phase in (b). (d) and (e) correspond to crossings at two different times. (h) is interesting because it includes turning and crossing at the same time. Finally, cases (i) and (j) are challenging because it is easy for the tracking algorithms to think that the two paths cross each other.

For three people (Table 3) there are lot more combinations possible. However, many of them are simple extensions of the two person cases, which we ignore. Cases (i), (j), (k), (n) correspond to crossings involving all three paths. Cases (g), (h) , (m) have two people turning without crossing paths. These cases are interesting when there is high error in location which can cause all three paths to be mis-interpreted. The case with five people (Table 3) (o) (p) contains even more complicated crossings and turns that enable us to better understand the performance of our algorithms.

In Tables 2,3 we indicate the performance of the algorithms using an **accuracy metric** defined as follows. For a given test case, we count the total number of points. We then find the number of points that are assigned to incorrect paths by the algorithm. The ratio of this value expressed as a percentage gives us a measure of the accuracy of the algorithm. Thus a value of 0 in the tables means a 100% accuracy whereas a value of 9 in the table means that 91% of the points were correctly assigned. In the tables, we have five values that correspond to the five error values of 0%,10%,20%,30% and 40% in locationing respectively.

5.2   Two Person Case

Let us first consider the three cases when two people walk in a straight line. The first case is when the two people walk in the same direction in phase (i.e., start with the same leg). The second case is same as the first one but the two people are out of phase, i.e., two people start walking with different legs. The last case is when the two people walk in opposite directions. As Table 2 shows, the out-of-phase case has a lower accuracy as compared with the in-phase case. The accuracy is affected somewhat by the error in foot position but the effect is quite small. As we would expect, accuracy is higher when the two people walk towards each other since there is little possibility of confusing the foot positions.

Let us now consider the case when paths cross. If two paths cross at different times for the two individuals, then the problem of identifying the paths correctly is trivial. This case is illustrated in Figure 6 where A and B's paths cross but at very different times. On the other hand, consider the situation shown in Figure 7 where we see two people crossing each other almost simultaneously. In the figure, A crosses B's path at time units $t_A : 3, 4$ while B crosses at time units $t_B : 1, 2$. We define this case as a crossing occurring at $(3, 1)$ where we pick the earliest times for A and B when their paths cross. In order to study how the algorithms performed, we varied t_B over a range of times. For each topology design, we also applied location estimation error ranging from 0% to 40%.

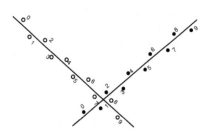

Fig. 6. Crossing path at $t_1 : 7, 8$ of the first object and $t_2 : 1, 2$ of the second object

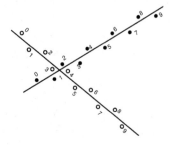

Fig. 7. Crossing path at $t_1 : 3, 4$ of the first object and $t_2 : 1, 2$ of the second object

Figure 7 shows the result of crossing at $t_A : 3, 4$ and from $t_B : 1, 2$ to $t_B : 6, 7$ with error between $0\% \sim 40\%$. For our algorithm, if the crossing points t_A and t_B differ by at least three then the algorithm will track the people correctly. In other words, A crosses at $t_A : j, k$, then if B crosses either before $t_B : j - 2, k - 2$ or after $t_B : j + 3, k + 3$, the algorithm will work correctly. The table also tells us how the location error affects the result. However, its effect is unpredictable.

Table 1. B crosses at different time instants while A crosses at time is 3,4

	0%	10%	20%	30%	40%			0%	10%	20%	30%	40%
crossing time 1,2	✓	✓	✓	✓	✓		*crossing time 4,5*	×	×	×	✓	×
crossing time 2,3	✓	✓	×	✓	✓		*crossing time 5,6*	×	×	×	✓	×
crossing time 3,4	×	×	✓	✓	✓		*crossing time 6,7*	✓	✓	✓	✓	✓

Table 2. Different combinations of two people walking. Number represents the error percentage from test case with 0% to 40% respectively.

0,0,0,0,2	0,0,0,1,3	0,0,0,0,0	0,0,0,6,20	0,0,0,0,0	0,0,0,3,2	0,0,0,0,5	0,0,0,0,0	0,0,0,0,0	0,0,0,5,11
(a)	(b)	(c)	(d)	(e)	(f)	(g)	(h)	(i)	(j)

In Table 2 (d), (e) we see the results for the cases when the different of crossing times are 2 and 3. The algorithm is more accurate when the crossing time is far apart while maintaining 94% accuracy for 30% locationing error in (d). It is interesting to compare the results for cases (d) and (h). We see that in (h), the algorithm works better than in (d) even though (h) includes a turn and the difference in crossing time is close to each other. The reason has to do with the angle between the two paths at the point where they cross. In (h), as we can see, the angle is much greater than in (d) thus, the algorithm has a smaller chance of mistaking points belonging to one person for the other.

Finally, consider cases (i) and (j), where two paths are at their closest to each other. Our algorithm has a reduced accuracy for (j) because sometimes the algorithm predicts an erroneous U shaped path for each person. However, the probability of this happening is small.

5.3   More Than Two People

Looking at Table 3 (d), (e), (i), (j), (n) we have different crossing scenarios. In general, our algorithm has very high accuracy. The only interesting case where it has some inaccuracy is (n). The reason for this is that the angle of incidence in the paths of the the bottom two people is very acute and thus the algorithm sometimes (for high location error) confuses the paths. We see similar results for Table 3 (q), (r), (s), (p). However, we note that cases (k) for three people and (e)

Table 3. Different combinations of three and five people walking. Number represents the error percentage from test case with 0% to 40% respectively.

0,0,0,4,3	0,0,0,8,17	0,0,0,0,0	0,0,0,6,2	0,0,0,7,2	0,0,0,0,0	0,0,0,0,0	0,0,0,0,0	0,0,0,0,3	0,0,0,1,1
(a)	(b)	(c)	(d)	(e)	(f)	(g)	(h)	(i)	(j)
0,0,0,0,0	0,0,0,0,0	0,0,0,0,1	0,0,8,5,5	0,0,0,0,1	0,0,0,1,16	0,0,1,4,3	0,0,0,0,4	0,0,0,0,0	0,0,2,6,10
(k)	(l)	(m)	(n)	(o)	(p)	(q)	(r)	(s)	(p)

for five people are similar in that we have people coming in different directions. In all these cases our algorithm still performs very well even for 40% location error.

6   Related Work

Tracking multiple people in a room is a challenging problem. Traditional approaches have included using cameras/ pattern matching and sound but are typically very expensive and inaccurate. In the past few years, many researchers have employed different new technologies to track, locate and identify multiple people. In paper [1], they track multiple people using acoustic energy sensors. Their approach is to use sound signatures to locate people. However, we don't know how well the algorithm performs as the number of people increase and with path crossings etc. Indeed, identifying path crossings and turns is the hardest path of tracking and it is unclear how their approach would work.

An alternative which many people pursue is using sensor networks such as [3] and [4]. These papers use sensors equipped with sound recognition and show a 95% accuracy. However, the research study is only limited to straight line walking and with up-to four people in the sensing area. In addition, since it is based on acoustic signatures, microphone-equipped sensors are used. The problem is that noise in the environment will affect the accuracy.

[6] and [8] both use similar techniques to recognize the motion of foot steps. The algorithm is based on foot signature recognition. Unfortunately this approach will not scale when presented with an unknown person. Furthermore, they use a special pressure sensor that provides direction of motion as well. These sensors require the tiles to be elevated which further reduces the utility. [2] and [5] also use similar pressure sensors under tiles to locate people. In addition, they both use RFIDs so that they can identify the number of people when there is confusion. [5] uses cameras with knowledge of the number of people for tracking.

7   Conclusions

We examine the problem of tracking people in indoor environments. Using in-expensive pressure sensors, we develop good placement strategies that allow us to locate foot positions accurately. Given these positions, our tracking algo-rithm can track multiple people with high accuracy even if they turn and their paths cross.

References

1. Liu, J., Chu, M., Liu, J., Reich, J., Zhao, F.: Distributed State Representation for Tracking Problems in Sensor Networks. In: Proc. of 3nd workshop on Information Processing in Sensor Networks (2004)
2. Mori, T., Suemasu, Y., Noguchi, H., Sato, T.: Multiple People Tracking by Inte-grating Distributed Floor Pressure Sensors and RFID System. In: Proceedings of IEEE International Conference on System Man and Cybernetics (2004)
3. Mechitov, K., Sundresh, S., Kwon, Y., Agha, G.: Cooperative Tracking with Binary-Detection Sensor Networks. In: Proceedings of the 1st international con-ference on Embedded networked sensor systems (2003)
4. Savarese, C., Rabaey, J.M., Beutel, J.: Locationing in Distributed Ad-hoc Wireless Sensor Networks. In: Proc. 2001 Int'l Conf. Acoustics, Speech, and Signal Process-ing (2001)
5. Kaddoura, Y., King, J., Helal, A.: Cost-Precision Tradeoffs in Unencumbered Floor-based Indoor Location Tracking. International Conference On Smart homes and health Telematic (2005)
6. Headon, R., Curwen, R.: Recognizing Movements from the Ground Reaction Force. In: Proceedings of the 2001 workshop on Perceptive user interfaces (2001)
7. Orr, R.J., Abowd, G.D.: The Smart Floor: A Mechanism for Natural User Identi-fication and Tracking. In: Proceedings of the 2000 Conference on Human Factors in Computing Systems (2000)
8. Addlesee, M.D., Jones, A.H., Livesey, F., Samaria, F.S.: The ORL Active Floor. IEEE Personal Communication (1997)
9. Stride Analysis.
Website: http://moon.ouhsc.edu/dthompso/gait/knmatics/stride.htm
10. Flexforce Pressure Sensor.
Website: http://www.tekscan.com/flexiforce/specs_flexiforce.html

Smart Mote-Based Medical System for Monitoring and Handling Medication Among Persons with Dementia

Victor Foo Siang Fook[1], Jhy Haur Tee[2], Kon Sang Yap[2], Aung Aung Phyo Wai[1], Jayachandran Maniyeri[1], Biswas Jit[1], and Peng Hin Lee[2]

[1] Institute for Infocomm Research
{Sffoo,Apwaung,Mjay,Biswas}@I2r.a-star.edu.sg
[2] Nanyang Technological University
{Dunforget,810924085345,Ephlee}@Ntu.edu.sg

Abstract. This paper presents a novel smart mote-based portable medical system which automatically monitors and handles medication among persons with dementia based on wireless multimodal sensors, actuators and mobile phone or PDA (Personal Digital Assistance) technology. In particular, we present the subtle design, implementation and deployment issues of monitoring the patient's behavior and providing adaptive assistive intervention such as prompts or reminders in the form of visual, audio or text cues to the patient for medical compliance. In addition, we develop mobile phone or PDA applications to provide a number of novel services to the caregivers that facilitate them in care-giving and to doctors for clinical assessment of dementia patients in a context enlightened fashion.

Keywords: Mote, Medication, Persons with dementia, Mobile phone or PDA.

1 Introduction

There is mounting worldwide interest to apply recent developments in context-aware systems, wireless sensor networks and mobile phone technology for healthcare. One area of focus is to develop activities-of-daily-living (ADL) behavior understanding system to facilitate caregiving and clinical assessment of demented elders within their homes. It is crucial for a physician to know whether the dementia patients are taking their daily medication at homes in order to prescribe the right dosage and to dispense correct advice on caring and coping to care-givers. This information is traditionally extracted from interviews with caregivers or even the patients themselves and suffers from serious problems of selective recall, knowledge gaps and inaccuracies. Hence, it is a huge challenge to physicians to promote patient adherence to the treatments.

However, failure in medical compliance will render the medical treatments ineffective and may lead to disastrous consequences. The father of medicine, Hippocrates, who already realized the importance of medical compliance more than twenty centuries ago once said: "The physician must not only be prepared to do what is right himself, but also make the patient cooperate". In this paper, we describe a novel smart mote-based portable medical system which automatically monitors and handles medication among persons with dementia at homes based on wireless

T. Okadome, T. Yamazaki, and M. Mokhtari (Eds.): ICOST 2007, LNCS 4541, pp. 54–62, 2007.

multimodal sensors and mobile phone or PDA technology to promote medication adherence. It provides assistive cues to patients in the form of prompts and reminders, and allows physicians or caregivers to monitor patient's medicine taking activity and obtain summarized behavioral reports from their PDA anytime, anywhere. Section 2 discusses the related works. Section 3 describes the design considerations and details of a smart medical system. Section 4 describes the PDA or mobile phone applications to the caregivers and doctors for caregiving and clinical assessment in a context aware fashion. Finally, section 5 concludes with a discussion of future works.

2 Related Works

Previously, many systems have been developed to support medication adherence. The assistive technology lab in University of Toronto proposes a medication reminding system [1] using context aware technology while ETH Zurich proposes a smart medicine cabinet [2] using passive RFID and Bluetooth-enabled active tags to monitor the contents of the box. The Lanchaster University [3] designs a device to support the management of medication in a community care environment, and the University of Ulster proposes a pill container [4] for medication. Recent works in Intel by Jay Lundell [5] proposes a smart context aware medication adherence system. Our work is similar to them in that we also use sensors, context aware systems and PDA or mobile phones to provide assistive cue to patients such as reminders for medical compliance. However, our work is different in some ways as we seek to provide a single holistic integrated portable smart medical system to collectively address and satisfy the different needs and perspectives of all stakeholders for medication compliance such as the patients, caregivers and doctors using the pervasive mote platform. In the long term, we hope to integrate more sophisticated behavior understanding system to provide holistic solutions to dementia patients beyond medication adherence using the popular mote platform.

3 Smart Medical System

In this section, we will describe our hardware and software design considerations of a smart medical system for monitoring and handling medication in dementia persons.

3.1 Design Considerations

We study the requirements from the perspectives of patients, caregivers and doctors, and also from the literature survey. In all cases, it requires the smart medical system to be safe but reliable enough to capture the medication taking behavior of patient. Good recognition rate must be achieved and false alarms should be minimized to improve the practicality of mass deployment of such a system. It should not be intrusive and not change the behaviors of the dementia patients. Furthermore, the devices should be adaptable to the changing environments such as cases in which temperature can change due to weather or artificial cooling and hence the sensors in the system must be temperature compensated. From patient's perspective, it should be portable so that a reminder can be sent to patients when they are not at home. The device should be easy to use or easily worn like a watch. Different video, audio and text cues may be provided for those with hearing problems, visual problems, etc and situated reminders should be sent as the dose might be missed due to sleeping.

From caregivers' perspective, if the patients decide not to or forget to take medication, at least a automated monitoring system or additional form of support such as situated alert being sent for the caregivers will bring some bearing to assist with their non-compliance. Likewise, from doctors' perspectives, the system should assist in the control of medication administration and drug therapy, and record the progress of the patient through a dedicated medication regime. In sum, the requirements of the smart medical system are two-fold: capture all the medication taking characteristic behavior relating to the dementia patients in a non-intrusive way, and intervene by processing and relaying information in a context-aware and distributed manner.

3.1.1 System Design

The smart medical system is designed to meet the above considerations and consists of a medication box with sensors, patient medication analyzer, central server and PDA or mobile phones. It is designed such that a smart medicine box can be connected to multiple related doctors and caregivers, and vice versa.

The medicine box periodically sends sensor readings wirelessly to the patient medication analyzer. Using a PDA, a doctor can first authorize himself through the central server, and connect to the patient medication analyzer to obtain behavioral reports of the patient in a distributed manner. This system is illustrated in Figure 1.

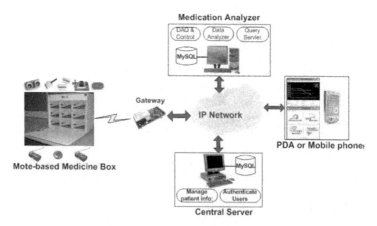

Fig. 1. The smart medical system

The smart medical system provides scalable monitoring provisioning and support standardized schemes for automated intervention management and activity planning. The detailed hardware and software components are described in the next sections.

3.1.2 Hardware

The wireless smart mote-based medicine box for medical adherence among persons with dementia is designed as shown in Figure 2.

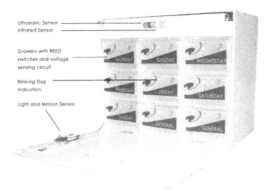

Fig. 2. The smart medicine box implemented with various sensor technologies

It utilizes multimodal sensors and actuators to monitor and assist patient in taking medicine. The smart medicine box consists of nine drawers – seven drawers for each day in a week which allows the patient to differentiate the medicine to be taken for a particular day and two general ones. A LED will blink to assist the patient in opening the right drawer. If a wrong compartment is opened, an error tone will be generated to alert the patient. In case the patient forgets to take the medicine, a reminder tone will be generated to remind the patient it is time to take his medicine. The details of the hardware components are described below:

- Motion Sensing Medicine Box Lid and Environmental Light Sensing

A mote with accelerometer and light sensor is attached to the main lid of the medicine box to sense whether it is opened or not. The lid is detected as open if the accelerometer sensor readings drop by an amount greater than 25 and if the light sensor detects a light intensity greater than 700 (maximum reading is 1024).

- Sensing Human Presence and Motion (PIR)

An external PIR sensor connected to the mote is used to sense the presence of the patient around the box. Output of the PIR sensor is connected to the ADC port on the sensor board. When motion is detected, this output readings will go below 600. An ultrasonic sound sensor is also used to determine the distance of patient from the medicine box. Signal from an external ultrasonic sensor powered by a 6V voltage source is fed into the ADC1 port of the MTS101CA sensor board on the mote. The ADC port interprets the ultrasonic sensor readings in a range from 0 to 1024. If main lid is closed, the distance to obstacle will be very small. When the lid is opened, the value will be very high (>200). Using a simple algorithm, the system reports the motion of patient in front when the readings are in the range from 20 to 100.

- Controlling LED Indicators

When the main lid is opened, LED on the drawer to be opened will start blinking. Since there are only limited motes and PWR output ports available, we design a circuit so that the 7 LEDs can be turned on or off efficiently. 3 output ports on the

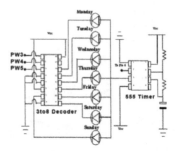

Fig. 3. LEDs control circuit schematic

sensor board of this mote, PW3, PW4, PW5, are used to control the LED indicators as shown in Figure 3. The 3 output voltages from the mote are connected to a 3 to 8 decoder. The outputs from the decoder are connected to the negative input of the 7 LEDs. The positive terminals of every LEDs are also connected to the output pin of a 555 timer, which generates a high and low voltage alternatively.

• Sensing Drawer Motion

A smaller mica2dot mote together with REED switches and a resistor circuit is used to achieve the above purpose. A circuit as shown in Figure 4 is implemented to sense the status of 9 drawers through 1 ADC port effectively. A reference voltage source supplies a series of 10 resistors in which 9 of the resistors are connected to 9 of the drawers and a resistor acts as a reference resistor. When a drawer is closed, a REED switch connected to it will be closed, and therefore its connected resistor will be short-circuited from the series. The voltage across the reference resistor will increase accordingly and can be used to detect whether the drawer is opened.

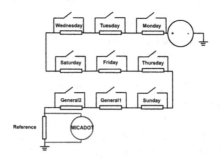

Fig. 4. Series resistor network to sense drawer status

• Tone Generation

Tones are generated by two output voltages to the buzzer/speaker when a mote receives a control message from the patient medication analyzer. The controlling circuit is depicted in Figure 5 below.

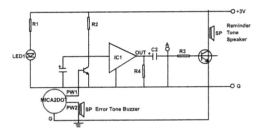

Fig. 5. Circuit to generate reminder and error tones

3.1.3 Software

We adopt a layered software architecture design to act as a platform for developers to build applications. Besides the NESC modules in the medication box, it consists of modules for the patient medication analyzer, central server and PDA/mobile phones.

3.1.3.1   Patient Medication Analyzer Modules. The analyzer modules collect sensor readings from the medicine box and store them into a database. Servlets running in tomcat are implemented to perform task requested by the PDA/mobile phone and also many tasks such as reply current alarm settings in database upon request, reply a text report for a particular day, etc. A Bayesian reasoning engine is also integrated for performing information fusion between multiple sensors. Using Bayesian network, we can infer the probability of an event objectively based on the data collected. As shown in Figure 6, a Bayesian network is used to calculate the probability that a patient is taking medicine at a particular instance. In the network, node "Taking-medicine" acts as the parent node for door status, drawer status and human detection nodes. Its status will directly affect the values of the three children nodes. For instance, if a patient is taking medicine, there is a higher probability that the lid is opened, correct drawer is

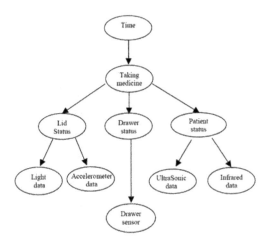

Fig. 6. Inferring Probability of Taking Medicine using Bayesian Network

pulled out, and a patient is detected in the proximity. Light and accelerometer data are used to infer lid status, and ultrasound and infrared data are used to sense whether the patient is nearby to the medicine box. Meanwhile, the drawer sensor data is used to indicate the drawer being pulled out. Time node in the Bayesian network acts as a prior probability node of patient taking medicine. The prior probabilities can be hour-based which means that each hour of a day has different prior probability. Every patient has different regular time on taking medicine, and hence the prior probabilities are not predetermined. A one-week time period samples are collected and used to train the prior probabilities. The trained prior probabilities will indicate the most likely as well as the least likely time of the patient to take medicine.

3.1.3.2   Central Server Modules. The central server is designed to manage the relationships among multiple medicine boxes and their respective caregivers or doctors. It is also used to perform authorization and maintain patient information that can be retrieved by the doctor.

3.1.3.3   Application Modules. Applications for the patients, caregivers and doctors are built on the platform and the details of the applications built on PDA/phone will be described in the next section.

4   PDA or Mobile Phone Applications

Java applications are developed for the doctors and caregivers to monitor the status of the medicine box or request behavioral report of a particular patient. It can be easily extended to act as reminders. The functionalities include authentication and patient selection, change alarm settings, request for live report, etc.

A login interface is presented for the user and once authorized by the central server, one has the options to view patient particulars, register a new patient into the system or update login information. The application will also show a menu which consists of four major features developed: 'Set Reminder', 'Live Report', 'Text Report' and 'Graph Report' as shown in Figure 7.

Fig. 7. Interface for caregivers or doctors

Feature 1: Set Reminder
Set Reminder feature allows one to remotely set 5 reminders for the medicine box. A request was first sent to the Analyzer to obtain current alarm settings, and then new settings will be sent to the servlets to update the database records.

Feature 2: Live Report
Live report feature allows user to view real-time status of the medicine box. Two options are available: Chart Mode or Image Mode as shown in Figure 8. For Chart mode, latest sensor readings are updated on the chart while Image Mode allows the user to gain pictorial view on the status of the medicine box in real time.

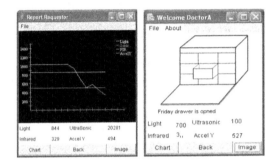

Fig. 8. Live Report – Chart View and Image View

Feature 3: Text and Graph Report
Text report shows summarized behavioral report of a patient in daily, weekly and monthly format generated based on Bayesian networks. Daily Report shows time slots when probability of taking medicine by patient is high by inferring it from the percentage of times the correct drawer is opened. Weekly and Monthly report conclude total number of times that the patient has taken or not taken medicine in weekly and monthly basis. Detailed Log for Daily Report shows peak periods when probability of medication activity is high, sensor modalities that contributed to the probability and the drawers opened at a particular period, and Graph Report displays full probability information of selected day in graphical form, as depicted in Figure 9.

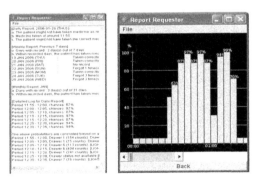

Fig. 9. Text and Graph Report

Experiments are conducted by students to simulate patient's behavior and the results are encouraging that we will try to deploy the system in patient's home for evaluation. While development is still in its early stages, our joint effort with a local hospital should see us achieving our objective of validating it in a real life setting.

5 Conclusion

We present a smart medical system for monitoring and handling medicine taking behaviors for dementia patients, caregivers and doctors. The use of multimodal sensors, actuators and PDA or mobile phone is the first step for us to promote medical compliance, and we are now furthering our work by including more sensor modalities such as pressure sensors, RFID, etcr. to enhance the recognition rate and reduce false alarm, and also working on adaptive interface such as LED status board for patients. The joint effort with a local hospital should see us achieving our long term objective of integrating more sophisticated behavior understanding system to provide holistic solutions to dementia patients beyond medication adherence using the mote platform.

References

1. Mihailidis, et al.: A context-aware medication reminding system: Preliminary design and development. Rehabilitation Engineering and Assistive Technology Society of North America, Atlanta, Georgia, CD-ROM Proceedings
2. Lampe, M., et al.: Advances in Pervasive Computing, Austrian Computer Society (OCG). Vienna, Austria (April 2004)
3. Kember, et al.: Designing Assistive Technologies for Medication Regimes in Care Settings. Universal Access in the Information Society 2, 235–242 (2003)
4. Nugent, et al.: Can Technology Improve Compliance to Medication, 3rd International Conference on Smart Homes and Health Telematics (2005)
5. Jay, L., et al.: Why Elders Forget to Take Their Meds: A Probe Study to Inform a Smart Reminding System, 4Th International Conference on Smart Homes and Health Telematics (2006)

Home Based Assistive Technologies for People with Mild Dementia

Chris Nugent[1], Maurice Mulvenna[1], Ferial Moelaert[2], Birgitta Bergvall-Kåreborn[3],
Franka Meiland[4], David Craig[5], Richard Davies[1], Annika Reinersmann[4],
Marike Hettinga[2], Anna-Lena Andersson[3], Rose-Marie Dröes[4],
and Johan E. Bengtsson[3]

[1] School of Computing and Mathematics, University of Ulster, Northern Ireland
{cd.nugent,md.mulvenna,rj.davies}@ulster.ac.uk
[2] Telematica Instituut, Enschede, The Netherlands
{Ferial.Moelaert,Marike.Hettinga}@telin.nl
[3] Luleå University of Technology, Luleå, Sweden
{Birgitta.Bergvall-Kareborn,Johan.E.Bengtsson}@ltu.se
Anna-Lena.Andersson@soc.lulea.se
[4] Depart. of Psychiatry/Alzheimer center, VU University medical center, The Netherlands
{fj.meiland,a.reinersmann,rm.droes}@vumc.nl
[5] Belfast City Hospital/Queen's University of Belfast, Northern Ireland
david.craig@qub.ac.uk

Abstract. Those suffering from mild dementia exhibit impairments of memory, thought and reasoning. It has been recognised that deployment of technological solutions to address such impairments may have a major positive impact on the quality of life and can be used to help perform daily life activities hence maintaining a level of independence. In this paper we present an overview of our current investigations into how technology can be used to improve the quality of life of the ageing person with mild dementia. Specifically, we detail the methodology adopted for our work, outline results attained from a series of workshops to identify user needs and finally present how these user needs have been mapped onto the design of home based assistive technologies.

Keywords: Assistive technologies, independent living, mild dementia, mobile devices, intelligent environments.

1 Introduction

Dementia is a progressive, disabling, chronic disease affecting 5% of all persons above 65 and over 40% of people over 90 [1]. In Europe approximately 1.9 million people experience mild dementia. Typical symptoms exhibited include impairments of memory, thought, perception, speech and reasoning. Early impairments in performing complex tasks lead to an inability to perform even the most basic functional activities such as washing and eating. In addition to these impairments it is also common to witness changes in personality, behaviour and psychological functioning, such as symptoms of depression, apathy and aggression.

Demographic changes mean that we can expect a rise in the number of ageing people and hence a rise in the numbers of people with mild dementia. The impact of

T. Okadome, T. Yamazaki, and M. Mokhtari (Eds.): ICOST 2007, LNCS 4541, pp. 63–69, 2007.

this increase of ageing people within the population will be long waiting lists for sheltered housing projects, homes for the elderly, nursing homes and other care facilities. The majority of people with dementia will have to 'survive' in their own homes. Nevertheless, it has been suggested that most ageing people prefer to stay at home as long as possible, even if they are at risk [2]. On the one hand this reduces the pressures on nursing homes and other similar types of care facilities, however, it increases the pressure on both formal and informal carers. This is further complicated through the generally appreciated fact that there is an increasing shortage of professional carers and in today's society there are fewer young people (for example younger family members of the person with dementia) to care for the ageing person.

In our work we aim to develop services for ageing people with mild dementia with a focus on the real needs and wants of such users. In addition, our developments propose to help address the societal problems which are witnessed as an adverse effect of those with mild dementia. We are investigating how technology can be used to improve the autonomy and the quality of life of ageing people with mild dementia and hence offer a means by which the person can remain in their own home for a longer period of time with an improved quality of life. A secondary impact of such a solution is anticipated to offer increased benefits in terms of relieving the burden for formal and informal carers.

2 Background

While there is some research and development in cognitive prosthetics, there are very few relevant tools, solutions or technologies specifically for people with mild dementia. In addition, although there is evidence reporting the unmet needs of persons with mild dementia, there is no currently available solution or research results which provides a complete solution to address the full suite of such needs. This has resulted in a number of studies addressing specific individual areas of concern developed more for the ageing population as opposed to specifically for those with mild dementia. For example, a number of studies have reported on the use of general memory aids which can be used by those suffering from memory problems and cognitive impairments [3, 4]. Such devices can be pre-programmed to offer audible and visual reminders throughout the day through a varying form of interfaces. Mobile phone based technologies have been reported as a potential means to promote social contact [5]. In such cases customisable solutions in the form of single button devices have been proposed. A number of solutions to support general daily life activities for the ageing have been reported. These have included, for example, medication management services, item locators and the offering of remote services to support healthcare provision [6]. Finally, solutions in the form of electronic tagging have been described as potentially successfully solutions to support the constant safety of a person within residential homes and the general community [7].

Taking into consideration the needs of people with mild dementia in conjunction with relevant technological solutions it has been possible to identify potentially innovative solutions to offer cognitive reinforcement. Specifically, the core scientific and technological objectives of our work aim to achieve a breakthrough in the development of a successful, user-validated cognitive prosthetic device with associated services for people with mild dementia. We will address this by focusing on the development of the following technological services:

- remotely configurable reminding functionality
- communication and interaction functionality
- supportive technology for performing activities of daily living
- anomaly detection and emergency contact

In the following Sections we outline the methodology by which we aim to realise these objectives, provide an insight into the results attained from a series of user workshops and present a technological overview of the home based assistive technologies to be developed within our work.

3 Methods

To address the scientific and technological objectives of our work we have adopted a user centred design approach which is based on a series of three iterations. Each iteration is composed of three generic phases. The first phase aims to assess the user needs, the second phase is concerned with the technical development of the user requirements in the form of a series of prototypes and the third and final phase is one of user testing and evaluation. The results from the final test and evaluation phase of each iteration are used in conjunction with the user requirements to provide the direction of required the technical re-development and refinement in ensuing iterations.

The starting point is to ascertain the functionality, performance and other requirements which are required to be fulfilled in order for the system to be deemed adequate for testing by those involved from the user perspective. Users can take two forms:

1. Priority target group – people with mild dementia, who require assistance with performing certain daily activities and who may benefit from use of the developed service in terms of greater quality of life.
2. Persons who support the priority target group – these are informal and/or professional carers who support and provide care for the user and will be responsible for configuring the system.

Both sets of users are seen as collaborators in the developmental process and will be asked to make comments on the performance, reliability, usefulness, safety factors, suitability or desirability of the developed service. Through a series of workshops, insights into the needs, desires and demands of the users with reference to key areas of cognitive reinforcement will be identified. Workshops will be held in three different sites across Europe. Results from the workshops will be translated into a set of functional requirements which will be used as the basis for the technical specification for the prototypes. Prototypes, once developed, will be tested in the same three sites where the workshops were held.

4 Results

Initially we identified four main areas of cognitive reinforcement within which we aimed to offer solutions to people with mild dementia, providing the ability to have greater actual and perceived autonomy and improved quality of life. The areas of cognitive reinforcement identified were as follows:

- remembering
- maintaining social contact
- performing daily life activities
- enhanced feelings of safety

To investigate the needs of users within these four areas, workshops were conducted in three different European sites; Amsterdam (The Netherlands), Lulea (Sweden), Belfast (Northern Ireland). During each workshop, interviews were conducted with people with mild dementia (n=17) and their carers (n=17) according to a common structured protocol and suite of questions. At least two members of the Project's research team where present during the workshops. One person was responsible for leading the discussions and interviews whilst the second person recorded details of the discussions.

Table 1 presents a summary overview of the findings, in terms of user requirements, from all three workshops. Further details of the workshops conducted can be found in [8].

Table 1. Summary of needs and wants following three workshops in Amsterdam (The Netherlands), Lulea (Sweden) and Belfast (Northern Ireland). For the technical development in each of the three planned iterations within the project, the user needs have been allocated to be addressed in either Field Trial #1, #2 or #3. User needs stated in italics within the table will be realised within the first Field Trial.

Area of Cognitive Reinforcement	Summary of required services/solutions from all three workshops for people with mild dementia (n=17) and their carers (n=17)
Remembering	Item locator (keys, mobile device)
	Means to remember names of person based on pictures of faces
	Reminding functionality for common activities for example appointments
	Support with remembering day and time
Maintaining Social Contact	Means to provide communication support with carer/family network
	Ability to have a video telephone call
	Picture dialing
Performing daily life activities	Support with daily activities associated with pleasure
	Control of household devices e.g. television, radio, planning activities
	Music playback
Enhanced feelings of safety	*Warnings for doors left open*
	Warnings for devices left on
	General: way to contact others in instances of emergency

The results indicate the preferences based on the aforementioned four areas of cognitive reinforcement. To support the development of the first iteration of the technical prototype the user requirements were ranked based on their technical feasibility, preferences at each workshop site and following a review of current ICT solutions which have already been proven useful for persons with mild dementia. This approach accommodated for the variation in both preferences and user needs recorded across each of the three workshop sites.

To support the further understanding of the user needs two scenarios were created with the goal of clarifying how the technology could be used in daily life and how they relate to the specific user needs.

Based on Table 1, a set of functional requirements may be identified which can then be mapped onto the technical specification for prototype development. Given the iterative nature of the proposed plan of work within the Project the functional requirements were allocated to either the first, second or third planned technical development stage and Field Trial evaluation. This approach also provides the flexibility, following each evaluation phase, to adjust or refine the functional requirements and technical specification.

Essentially, to realise user needs as identified during the workshops, the system in its entirety will be comprised of four main components:

- *Mobile based cognitive prosthetic.* This will be the main interface between the system as a whole and the person with mild dementia. The device will act as a means to deliver reminder messages and can also be used to control various home devices in addition to offering a form of voice communications. For the first iteration of Field Trials the HTC P3300 smartphone will be used. It is also planned to 'tag' the device with an RFID tag. The purpose of including this tag will be to monitor if the person is about to leave the home without the device and issue a warning if this situation arises.

- *Home based hub.* This device will act as the main gateway within the person's home and will communicate with the cognitive prosthetic, the services outside the home and all devices within the home environment. The hub will provide a limited local means of data processing to avoid continually transmitting data to the server. The home hub will be controlled via a touch screen. Through a series of icons the user will be able to select from the following available services; picture dialling, locating the cognitive prosthetic and a dedicated 'alarm button' which will activate a direct alarm message being sent to a nominated carer.

- *Sensorised environment.* The home will be required to have a number of devices specifically equipped with sensory and control elements for example doors, television, radio etc. to permit assessment of the living environment. In the first Field Trial this will be limited to the control of mains power facilitating the turning 'on' and 'off' of devices such as the television and radio. In addition, the front door to the home will be equipped with sensors to monitor whether the door has been closed and also if it has been locked.

- *Web based server.* This component of the system will support the management of the person with dementia's reminding regimen. This will be used by both formal and informal carers. The system will have the ability to record all events associated with the person with dementia and provide a means for their carers to monitor these. In addition, the system will provide the ability for the carers to enter various reminders and functionality for the cognitive prosthetic for example details of appointments. The backend system will be developed in PHP/PostgresSQL.

Figure 1 provides an overview of the technical components required to address the user needs. Where possible developments will be based on the usage off-the-shelf hardware to reduce development times.

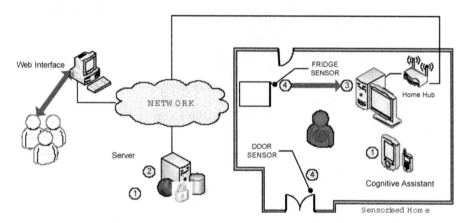

Fig. 1. Overview of technical components of home based assistive technologies driven through assessment of results from first iteration of requirement gathering workshops

5 Conclusion

The increasing numbers of ageing people with mild dementia has been identified as an area which could benefit from the deployment of innovative technological solutions. Such solutions will have the ability to promote the independence of those with mild dementia, improve their quality of life and extend the period of time they can remain living within their own home. Within our current work we aim to develop services and solutions for ageing people with mild dementia with a focus on the real needs and wants of such users.

In this article we have outlined the methodology within which we will conduct our research. We are currently in the first iteration of our approach. We have presented a summary of the results from our first three workshops which have been used to identify the real needs of the users (both people with mild dementia and their carers) and have shown how we have mapped these functional requirements onto a technical model which will meet the needs of all those involved.

Acknowledgements

This work has been partially funded by the European Commission under grant 034025.

References

1. Fratiglioni, L., Launer, L.J., Anderson, K., et al.: Incidence of dementia and major subtypes in Europe: A collaborative study of population-based cohorts. Neurol Dis. Elderly Res. Group, Neurology 54, 10–15 (2000)
2. Cook, D., Das, S.K.: How smart are our environments? An updated look at the state of the art. Journal of Pervasive and Mobile Computing (in press) (2007)
3. Holthe, T., Hage, I., Bjorneby, S.: What day is it today? Journal of Dementia Care. 7, 26–27 (1999)
4. Oriani, M., et al.: An electronic memory aid to support prospective memory in patients in the early stages of alzheimer's disease: a pilot study. Aging & Mental health 7, 22–27 (2004)
5. Lekeu, F., Mojtasik, V., van der Linden, M., Salmon, E.: Training early Alzheimer patients to use a mobile phone. Acta. Neurol. Belg. 102, 114–121 (2002)
6. TeleCARE: A muli-agent telesupervision system for elderly care
http://www.uninova.pt/ telecare/
7. Miskelly, F.: A novel system of electronic tagging in patients with dementia and wandering. Age. and Ageing 33, 304–306 (2004)
8. Meiland, F.J.M., Craig, D., Molaert, F., Mulvenna, M., Nugent, C., Scully, T., Bengtsson, J., Droes, R.M.: Helping people with mild dementia navigate their day. iN: Proc. ICMCC, Amsterdam (in press) (2007)

Roadmap-Based Collision-Free Motion Planning for Multiple Moving Agents in a Smart Home Environment

Jae Byung Park and Beom Hee Lee

School of Electrical Engineering and Computer Sciences
Seoul National University, Seoul, Korea
{pjb0922,bhlee}@snu.ac.kr

Abstract. This paper proposes an approach to collision-free navigation for multiple moving agents in a smart home environment using the visibility map which is one of road maps. The visibility map plans collision-free paths of agents against stationary obstacles. The collision-free paths are represented by Hermite curves taking into consideration smoothness of the path and its first derivative. For collision-free motion planning among multiple moving agents, the collision map scheme proposed to effectively analyze collisions between two robots is employed. According to the result of the collision analysis, time scheduling is applied for realizing collision avoidance. Simulation results show the effectiveness and feasibility of the proposed approach in collision-free navigation of multiple moving agents.

Keywords: Collision-free Motion Planning, Multiple Agents, Visibility Map.

1 Introduction

In forthcoming ubiquitous era, multiple moving agents such as a mobile robot, an unmanned aerial vehicle (UAV), and an autonomous underwater vehicle (AUV) will coexist in U-spaces like U-Home, U-Factory, and U-City. In such U-spaces, navigation of multiple moving agents is a crucial issue. However, collision-free motion planning of multiple agents in a dynamic environment turns out to have NP-hard problems and not to be solved mathematically [1]. Thus, the previous researches has been attempted to solve the combined problem by a numerical method.

In this paper, we solve the problem for collision avoidance of multiple moving agents with curve-shaped paths using the collision map scheme [2]. First, we have to acquire a collision-free path against stationary obstacles from a starting position to a goal position of an agent. There are many algorithms for path planning such as visibility graph search [3], Voronoi diagram [4], free space search [5], potential field method [6], and others. In this paper, we get a set of via-points of the collision-free path using the visibility map. Next, each edge between two adjacent via-points of the collision-free path is modeled as Hermite curves. The Hermite curve consists of starting and ending points, and first derivatives at those points. Therefore, each path consisting of several Hermite curves can be easily established to satisfy the smoothness of the path and its first derivative. The first derivative of the path coincides with the heading direction of the agent. Since an actual agent is not able to

T. Okadome, T. Yamazaki, and M. Mokhtari (Eds.): ICOST 2007, LNCS 4541, pp. 70–80, 2007.

change its heading direction abruptly, the first derivative smoothness of the path should be satisfied. For collision avoidance among multiple moving agents with the obtained collision-free path against stationary obstacles, the extended collision map scheme for multiple agents [7] is employed. We first build a collision map with the obtained collision-free paths. A time delay method for time scheduling is applied for collision avoidance among multiple moving agents.

In Section 2, the problem formulation for collision avoidance of multiple moving agents is presented. The visibility map for finding a collision-free path against stationary obstacles and the Hermite curve for modeling the curve-shaped path are introduced. Also, several assumptions are stated. In Section 3, the original collision map scheme for two agents with the straight line path is generalized for two agents with curve-shaped paths. Then, in Section 4, a time delay method with the collision map is suggested for collision avoidance of multiple moving agents. In Section 5, simulation results are presented to demonstrate the effectiveness and feasibility of the proposed approach. Finally, concluding remarks are presented in Section 6.

2 Problem Formulation

For collision-free path planning against stationary obstacles, the visibility map is introduced. The collision-free path is modeled as a curve path using sequential Hermite curves, taking into consideration smoothness of the path and its first derivative.

2.1 Visibility Map

The standard visibility map is defined in a two-dimensional polygonal configuration space [7]. The nodes of the visibility map shares an edge if they are within line of sight of each other, where the nodes include the vertices of obstacle polygons in the agent workspace and the starting and goal positions. Let $V=\{v_1,...,v_n\}$ be the set of nodes of the visibility graph where v_i for $i=1,2,...,n$ is a column vector. If $v_i \in V$ and $v_j \in V$ are mutually visible to each other, there is no intersection between edge $\overline{v_i v_j}$ for a possible path segment and edge $\overline{v_p v_q}$ where $v_p \in V$ and $v_q \in V$ are two adjacent vertices of an obstacle polygon. For $r=[0,1]$ and $s=[0,1]$, two edges can be represented as $\overline{v_i v_j} = v_i + r(v_j - v_i)$ and $\overline{v_p v_q} = v_p + s(v_q - v_p)$, respectively. Therefore, equation $\overline{v_i v_j} = \overline{v_p v_q}$ can be represented as a matrix form as follows:

$$\left[v_j - v_i \quad -(v_q - v_p)\right]\begin{bmatrix} r \\ s \end{bmatrix} = \left[v_p - v_i\right]. \tag{1}$$

If the determinant of the matrix $[v_j\text{-}v_i -(v_q\text{-}v_p)]$ exists and both solutions r and s are within range $[0,1]$, $\overline{v_i v_j}$ and $\overline{v_p v_q}$ have an intersection between them. Thus, $\overline{v_i v_j}$ is not included in the set of edges of the visibility graph. In the visibility map, a weight of each edge is determined by the traveled length of the edge. In this paper, the shortest path from the starting position to the goal position is obtained from the weighted visibility map using the dijkstra's algorithm [8].

2.2 Hermite Curve Path Modeling

Let $C=\{v_1,v_2,...,v_m\}$ be a set of via-points of the collision-free path obtained by the visibility map, where $v_i \in V$ for $i=1,2,...,m$. For two adjacent points v_i and v_{i+1}, let e_i be an edge between v_i and v_{i+1}. Edge e_i is formed by the Hermite curve as follows:

$$e_i(t)=(2t^3-3t^2+1)\cdot v_i+(-2t^3+3t^2)\cdot v_{i+1}+(t^3-2t^2+t)\cdot v'_i+(t^3-t^2)\cdot v'_{i+1}, \tag{2}$$

where $i=1,2,...,m$-1 and $t=[0,1]$. Also, v'_i and v'_{i+1} are the first derivatives at points, v_i and v_{i+1}, respectively. That is, the Hermite curve consists of two points, v_i and v_{i+1}, and the first derivatives, v'_i and v'_{i+1}, at those points, respectively. For curve path modeling, the first derivatives, v'_1 and v'_m, at starting and goal points, v_1 and v_m, are determined by desired heading directions of the agent at those points, respectively. Except starting and goal points, we determine v'_i tangent to vector $\overline{v_{i-1}v_{i+1}}$ for $i=1,2,...,m$-1. Since e_{i-1} and e_i are continuous at via-point v_i, and the first derivative of e_{i-1} is same as that of e_i at that point as shown in Fig. 1, the resultant curve path is differentiable. This is a very important condition of a practical path since an actual agent cannot abruptly turn its heading direction because of nonholonomic constraints of it. Consequently, we can obtain a smooth path for an actual agent.

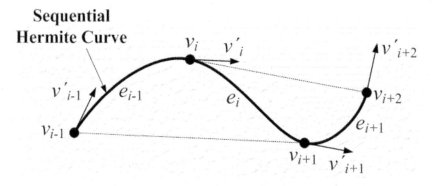

Fig. 1. Hermite curve model for a smooth path of an agent

2.3 Basic Assumptions

We solve collision avoidance for N mobile agents which move together in common workspace. Thus, collision among mobile agents can occur although each agent has a collision-free path against stationary obstacles generated by the visibility map. The moving obstacle avoidance is the NP-hard problem [9], [10]. For solving this problem, assumptions of our study are stated as follows:

Assumptions:
1) In a smart home environment, locations of agents are given.
2) Agents' paths are determined by the visibility map.

3) Agents with their trajectories follow their path within a tolerable error.
4) Collisions among agents do not occur at starting and goal positions.
5) Each agent is modeled as a circle.
6) Priorities of agents are predetermined.

3 Collision Map Analysis for Curve-Shaped Paths

The collision map concept was first suggested for collision detection and avoidance of two moving agents like robots with a straight line path [2]. In this paper, we deal with the collision map concept for a curve-shaped path modeled by the Hermite curve. The collision map bases on path and trajectory information of two moving agents. From the trajectory information, the traveled length $\lambda(k) \cdot l_{total}$ and the corresponding servo time instant $t=kT$ of an agent can be obtained along the path, where $0 \leq \lambda(k) \leq 1$, T is the sample time interval, and l_{total} is the total length of the path. As shown in Fig. 2, let $C_1=\{a_1,a_2,...,a_{m1}\}$ and $C_2=\{b_1,b_2,...,b_{m2}\}$ be the via-point sets of two paths of agents A_1 and A_2, respectively. Let $E_1=\{e_{1,1}(t_1),\ e_{1,2}(t_2),...,\ e_{1,m1-1}(t_{m1-1})\}$ and $E_2=\{e_{2,1}(s_1),\ e_{2,2}(s_2),...,\ e_{2,m2-1}(s_{m2-1})\}$ be the sets of edges between each two adjacent points of C_1 and C_2, respectively. As described in Section 2, each edge of E_1 and E_2 are formed by the Hermite curves in $t_u=[0,1]$ and $s_v=[0,1]$ for $u=1,2,...,m1-1$ and $v=1,2,...,m2-1$. Therefore, we can denote total length l_{total} of the A_2 path as follows:

$$l_{total} = \sum_{v=1}^{m2-1} \int_0^1 \sqrt{\left\{\frac{dx_{2,v}(s_v)}{ds_v}\right\}^2 + \left\{\frac{dy_{2,v}(s_v)}{ds_v}\right\}^2} \, ds_v \qquad (3)$$

where point $(dx_{2,v}(s_v), dy_{2,v}(s_v))$ is a point on the vth edge $e_{2,v}(s_v)$. If A_2 at time k is on the point $e_{2,v'}(s')$, the traveled length $\lambda(k) \cdot l_{total}$ of A_2 until time k can be computed as the sum of the total lengths from the first edge to the $(v'-1)$th edge and the partial length of the v'th edge from $s_{v'}=0$ to $s_{v'}=s'$ as follows:

$$\lambda(k) \cdot l_{total} = \sum_{v=1}^{v'-1} \int_0^1 \sqrt{\left\{\frac{dx_{2,v}(s_v)}{ds_v}\right\}^2 + \left\{\frac{dy_{2,v}(s_v)}{ds_v}\right\}^2} \, ds_v$$
$$+ \int_0^{s'} \sqrt{\left\{\frac{dx_{2,v'}(s_{v'})}{ds_{v'}}\right\}^2 + \left\{\frac{dy_{2,v'}(s_{v'})}{ds_{v'}}\right\}^2} \, ds_{v'} \qquad (4)$$

Using l_{total}, $\lambda(k)$ can be obtained by (4). Using trajectory information of A_2, $\lambda(k)$ is easily converted to k.

The radii of two moving agents, A_1 and A_2, are denoted as r_1 and r_2, respectively, since each agent is modeled as a circle. For the sake of simplifying the problem, let the radius of A_1 be r_1+r_2. Thus, we can consider A_2 as a point. In Fig. 2, A_1 has to move from a_1 to a_{m1} and A_2 has to move from b_1 to b_{m2}. Let $\mathbf{p}_1(k)$ be a point vector on the A_1 path at time k, where $\mathbf{p}_1(k)$ is obtained by the Hermite curve equation for edge

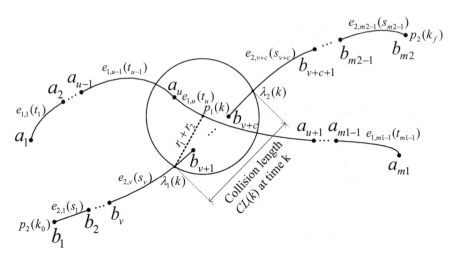

Fig. 2. Geometric analysis of collisions between two agents A_1 and A_2 with curve-shaped paths

$e_{1,u}(t_u)$. Since the collisions between A_1 and A_2 occur at time k when the distance between $\mathbf{p}_1(k)$ and the A_2 path is less than or equal to r_1+r_2, we solve the following equation:

$$(r_1+r_2)^2=\| \mathbf{p}_1(k)\text{-}\mathbf{e}_{2,v}(s_v) \|^2, \tag{5}$$

where $v=1,2,...,m2\text{-}1$ and $\mathbf{e}_{2,v}(s_v)$ is a point vector for a point on edge $e_{2,v}(s_v)$. In (5), we assume that s_v is equally quantized by $\varDelta s$ like $s_v=k_v\varDelta s$, and define an edge between each two adjacent points of the quantized edge $\mathbf{e}_{2,v}(k_v)$ as a straight line. The k_vth straight line equation of the vth edge $\mathbf{e}_{2,v}(k_v)$ is denoted as

$$\mathbf{p}_2= \mathbf{e}_{2,v}(k_v)+\gamma_{v,kv}(k)\cdot(\mathbf{e}_{2,v}(k_v+1)\text{-}\mathbf{e}_{2,v}(k_v)), \tag{6}$$

where $0\leq\gamma_{v,kv}(k)\leq1$. The equation (5) can be rewritten for each line segment as

$$(r_1+r_2)^2=\| \mathbf{p}_1(k)\text{-}\mathbf{p}_2 \|^2. \tag{7}$$

Substituting (6) for $\mathbf{p}_2$, we have

$$(r_1+r_2)^2=\{\mathbf{p}_1(k)\text{- }\mathbf{e}_{2,v}(k_v)+\gamma_{v,kv}(k)\cdot(\mathbf{e}_{2,v}(k_v+1)\text{-}\mathbf{e}_{2,v}(k_v))\}^T$$
$$\cdot\{\mathbf{p}_1(k)\text{- }\mathbf{e}_{2,v}(k_v)+\gamma_{v,kv}(k)\cdot(\mathbf{e}_{2,v}(k_v+1)\text{-}\mathbf{e}_{2,v}(k_v))\}. \tag{8}$$

More explicitly,

$$(r_1+r_2)^2=\| \mathbf{p}_1(k)\text{- }\mathbf{e}_{2,v}(k_v) \|^2\text{-}2\gamma_{v,kv}(k)\cdot(\mathbf{p}_1(k)\text{-}\mathbf{e}_{2,v}(k_v))^T$$
$$\cdot(\mathbf{e}_{2,v}(k_v+1)\text{-}\mathbf{e}_{2,v}(k_v))+\gamma_{v,kv}(k)^2\cdot\| \mathbf{e}_{2,v}(k_v+1)\text{-}\mathbf{e}_{2,v}(k_v) \|^2, \tag{9}$$

Eq. (9) is a quadratic equation in $\gamma_{v,kv}(k)$, which can be solved easily. From the trajectory information of A_2, we can denote the real roots of (9) as s_v for the vth curve like pair (v, s_v). Using the pair (v, s_v), we can obtain the corresponding $\lambda(k)$ from (4). Let $CS=\{\lambda_1(k), \lambda_2(k),..., \lambda_r(k)\}$ be the set of r ordered roots over the A_2 path, where one double real root is considered as two real roots and sequentially inserted to the set CS. Thus, r is always even. In this case, the collisions between two agents can occur at time k under the *collision length* with range from $\lambda_i(k) \cdot l_{total}$ to $\lambda_{i+1}(k) \cdot l_{total}$, where i is an odd number from 1 to r-1. We define the union of collision lengths at the collection of servo time instants, which determines the points on the A_1 path, as *collision regions* depicted in Fig. 3. Consequently, for collision avoidance of two agents, the traveled length versus servo time curve (TLVSTC) for A_2 denoted as TLVSTC$_2$ should not enter any collision region. Since it is difficult to describe the geometric shape of collision regions exactly and analytically, *collision boxes* are established to approximate collision regions. The starting time and ending time of the collision region are determined by only one real double root.

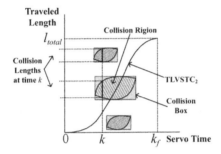

Fig. 3. Collision map concept with three collision regions for the curve-shaped path

4   Collision Avoidance for Multiple Moving Agents

Using the collision map and the time scheduling concepts, this section presents a method for collision-free motion planning of multiple moving agents. In this paper, a time delay method is employed as the time scheduling method. Fig. 4 (a) shows the effect of time delay on the collision map for two agents where it assumes that the priority of agent A_1 is higher than that of A_2. In the figure, the original TLVSTC$_2$ crosses the collision box as a potential collision region with A_1. Therefore, A_2 must be collided with A_1. However, after applying time delay to the original TLVSTC$_2$, the time-shifted TLVSTC$_2$ does not cross the collision box. That is, A_2 does not collide with A_1. So far, we dealt with collision avoidance between two agents. From now, we will consider collision avoidance among N agents based on the extended collision map [7], where the priorities among N agents are predetermined as $A_1 > A_2 > \ldots > A_N$. As shown in Fig. 4 (b), the delay time for the kth agent A_k can be determined, taking into consideration collision box CB$_{i,k}$ for $i=1,2,\ldots,k$-1, where CB$_{i,k}$ is a collision box of A_k with respect to A_i. That is, TLVSTC$_k$ is delayed until TLVSTC$_k$ does not cross any collision box. The time delay algorithm for collision avoidance among N moving agents is described in Table 1.

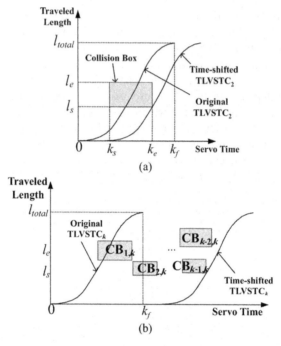

Fig. 4. Time scheduling on the collision map for collision avoidance. (a) Time delay for A_2 (k_s, k_e: Collision starting time and ending time, l_s and l_e: Collision starting point and ending point), and (b) Time delay for the kth agent A_k ($CB_{i,k}$: Collision box of A_k with respect to A_i for $i=1,2,\ldots,k-1$).

Table 1. Time delay algorithm for N moving agents

Step1	$k=1$	
Step2	Update collision map of A_{k+1} with respect to A_i for $1 \le i \le k$	
Step3	**IF**	$TLVSTC_{k+1}$ crosses collision box $CB_{i,k+1}$ with respect to A_i for $1 \le i \le k$
	THEN	Apply the time delay for collision avoidance to $TLVSTC_{k+1}$
Step4	**IF**	$k=N-1$
	THEN	Finish
	ELSE	$k=k+1$, go to Step2

5 Simulation Results

For verifying the suggested method, we carried out a simulation for collision avoidance among ten agents. Each agent A_i for $i=1,2,\ldots10$ where index i determines the priority of the agent. That is, A_1 and A_{10} are agents with the highest and lowest priorities, respectively. Simulation parameters are described in Table 2. For given starting and goal positions of ten agents, collision-free paths against five polygonal obstacles were obtained by the visibility map as shown in Fig. 5. The non-smooth paths with several straight lines in Fig. 5 (a) were modeled by the Hermite curves as smooth paths in Fig. 5 (b).

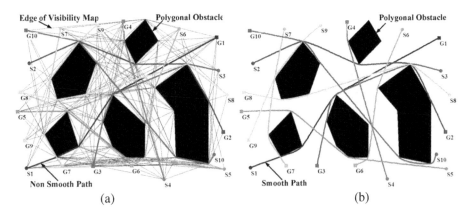

Fig. 5. Collision-free paths for ten agents from Sn to Gn for $n=1,2,...,10$. (a) Original non-smooth paths from the visibility map and (b) Smooth paths modeled by the Hermite curves.

Table 2. Simulation parameters for polygonal obstacles and agents

Parameters		Descriptions
O_1={(190,540), (143,456), (231,368), (275,500)}		
O_2={(456,543), (406,500), (368,406), (406,306), (531,418), (531,543)}		Five polygonal obstacle sets with vertex pairs as their elements where the unit is *cm*. There are two rectangle O_1 and O_5, one pentagon O_4, and two hexagons O_2 and O_3.
O_3={(650,543), (643,368), (568,281), (693,193), (775,331), (775,568)}		
O_4={(193,325), (168,243), (275,106), (343,185), (300,306)}		
O_5={(493,193), (450,106), (531,25), (581,106)}		
r_i=20 *cm* for i=1,2,...10		Radii of ten agents.
A_1: v_{max}=30 , a_c=2	A_6: v_{max}=20, a_c=4	Maximum velocities and accelerations of ten agents where each agent has a trapezoidal velocity profile. Units of the maximum velocity and acceleration are *cm/s* and *cm/s²*.
A_2: v_{max}=20, a_c=2	A_7: v_{max}=30, a_c=2	
A_3: v_{max}=30, a_c=2	A_8: v_{max}=30, a_c=1	
A_4: v_{max}=20, a_c=1	A_9: v_{max}=20, a_c=2	
A_5: v_{max}=30, a_c=3	A_{10}: v_{max}=15, a_c=3	
S1(75,581)	G1(806,95)	
S2(84,188)	G2(832,449)	
S3(813,219)	G3(332,574)	
S4(605,625)	G4(447,33)	
S5(836,588)	G5(40,369)	Starting and goal positions of ten agents in *cm*.
S6(662,62)	G6(480,573)	
S7(205,59)	G7(213,574)	
S8(847,310)	G8(49,297)	
S9(345,45)	G9(68,477)	
S10(797,533)	G10(71,65)	

While ten agents were moving along their paths, the minimum distance $D_{i,j}$ between each two agents A_i and A_j for i,j=1,2,...10 changed as shown in Fig. 6 (a). If $D_{i,j}$<0, two agent A_i and A_j collide with each other. The solid lines in Fig. 6 (a)

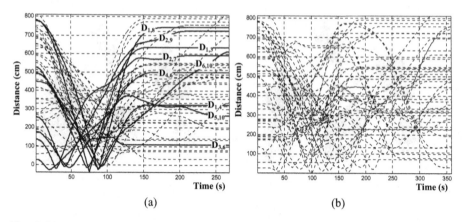

Fig. 6. Distance $D_{i,j}$ between two agents A_i and A_j for $i,j=1,2,...,10$ and $i \neq j$, where a collision between A_i and A_j occurs for $D_{i,j}<0$. (a) Before time scheduling, $D_{1,3}$, $D_{1,4}$, $D_{1,8}$, $D_{2,7}$, $D_{2,9}$, $D_{3,6}$, $D_{4,6}$, $D_{5,10}$ and $D_{6,10}$ depicted by solid lines are below 0 and (b) After time scheduling, no collision occurs, where distances depicted by dotted lines are above 0.

indicate the distances between two colliding agents with each other. The dotted lines indicate no collision between two agents. After time scheduling with the proposed collision map scheme, the distances between each two agents were always positive as shown in Fig. 6 (b). That is, there was no collision among all ten agents.

Time scheduling for collision avoidance among ten agents was conducted by using collision maps as shown in Fig. 7, where $CB_{i,j}$ is a collision box of A_j with respect to A_i and $TLVSTC_j$ is the traveled length versus servo time curve (TLVSTC) of A_j. In this case, A_1 just moved with its given velocity trajectory because A_1 is the agent with the highest priority. In Fig. 7 (a), the original $TLVSTC_2$ crossed $CB_{1,2}$. For collision avoidance with A_1, the original $TLVSTC_2$ should be time-shifted because the priority of A_1 was higher than that of A_2. As a result of time-shifting, the time-shifted $TLVSTC_2$ without collision with A_1 was determined. In Fig. 7 (b), the original $TLVSTC_3$ was time-shifted not to cross both $CB_{1,3}$ and $CB_{2,3}$ because the priorities of A_1 and A_2 were higher than that of A_3. In this case, $CB_{2,3}$ was determined by the time-shifted $TLVSTC_2$ as shown in Fig. 7 (a). Next, only three collision boxes $CB_{1,7}$, $CB_{2,7}$ and $CB_{5,7}$ are shown in Fig. 7 (c), because A_7 collides with only A_1, A_2 and A_5. In this case, the time-shift for A_7 did not needed because the original $TLVSTC_7$ did not cross any collision box. Collision boxes were also determined by the time-shifted TLVSTCs of A_i for $i=1,2,...,6$. In the case of A_{10} with the lowest priority, nine collision boxes with respect to A_i for $i=1,2,...9$ are shown in Fig. 7 (d). The original $TLVSTC_{10}$ was time-shifted with respect to $CB_{5,10}$. The resultant time-shifted $TLVSTC_{10}$ was determined between two groups of collision boxes. By this way, TLVSTCs for nine agents except A_1 were determined for avoiding collision with other agents.

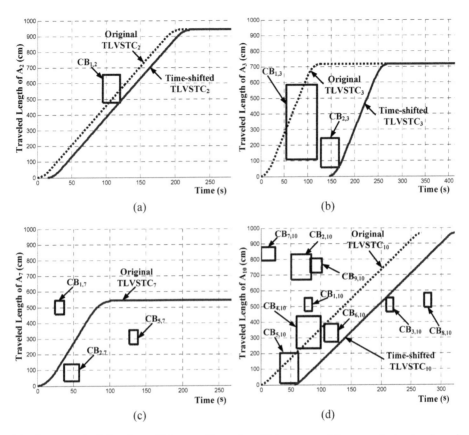

Fig. 7. Time scheduling for collision avoidance in collision maps of four agents A_2, A_3, A_7 and A_{10}, where $CB_{i,j}$ is collision box between A_i and A_j for i,j=1,2,...10. (a) Time delay of A_2 with respect to $CB_{1,2}$, (b) Time delay of A_3 with respect to $CB_{2,3}$, where $CB_{2,3}$ was determined by the time-shifted trajectory of A_2 in (b), (c) No time delay of A_7 since the original TLVSTC$_7$ did not cross collision boxes, and (d) Time delay of A_{10} with respect to $CB_{5,10}$ where the time-shifted TLVSTC$_{10}$ was determined between two groups of collision boxes.

6 Conclusions

Collision-free navigation of multiple moving agents in a smart home environment was studied using the collision map scheme. The collision-free path against stationary obstacles from a starting position to a goal position was obtained by the visibility map, and the curve-shaped path was modeled by the Hermite curve considering the smoothness of the path and its first derivative. Then, the collision map was analyzed for the curve-shaped path, and used for collision avoidance among multiple moving agents through time scheduling. The simulation results showed the feasibility of the proposed method.

Acknowledgments. This works was supported in part by the SRC/ERC of MOST/KOSEF under Grant R11-1999-008, and in part by the ASRI and BK21 at Seoul National University.

References

1. Akella, A., Hutchinson, S.: Coordinating the motions of multiple robots with specified trajectories. IEEE International Conference on Robotics and Automation, pp. 624–631 (2002)
2. Lee, B.H., Lee, C.S.G.: Collision-free motion planning of two robots: IEEE Transactions on Systems, Man and Cybernetics, vol. SMC-17(1) (1987)
3. Lozano-Perez, T., Wesley, M.A.: An algorithm for planning collision-free paths among polyhedral obstacles. Comm. ACM. 22(10), 560–570 (1979)
4. Mahkovic, R., Slivnik, T.: Constructing the Generalized Local Voronoi Diagram from Laser Range Scanner Data. IEEE Transactions on Systems, Man, and Cybernetics-Part. A: Systems and Humans 30(6), 710–719 (2000)
5. Meyer, W., Benedict, P.: Path planning and the geometry of joint space obstacles. IEEE International Conference on Robotics and Automation, pp. 215–219 (1988)
6. Khatib, O.: Real-Time obstacle avoidance for manipulators and mobile robots. Journal of Robotic Research 5(1), 90–98 (1986)
7. Choset, H., et al.: Principles of Robot Motion: Theory, Algorithms, and Implementations. The MIT Press, Cambridge (2005)
8. Cormen, T., et al.: Introduction to Algorithms. The MIT Press, Cambridge (2001)
9. Fujimura, K.: Motion Planning in Dynamic Environment. In: Computer Science Workbench, Springer-Verlag, Heidelberg (1991)
10. Hopcroft, J.E., Schwartz, J.T., Sharir, M.: On the complexity of motion planning for multiple independent objects; PSPACE-Hardness of the "Warehouseman's Problem. Journal of Robotics Research 3(4), 76–88 (1984)
11. Ji, S.H., Choi, J.S., Lee, B.H.: A Computational Interactive Approach to Multi-agent Motion Planning. International Journal of Control, Automation, and Systems (in press) (2007)

Peripheral Telecommunications:
Supporting Distributed Awareness and Seamless
Transitions to the Foreground

Yosuke Kinoe[1] and Jeremy R. Cooperstock[2]

[1] Faculty of Intercultural Communication, Hosei University,
2-17-1, Fujimi, Chiyoda, Tokyo 102-8160, Japan
kinoe@hosei.ac.jp
[2] Centre for Intelligent Machines, McGill University,
Montreal, Quebec H3A 2A7, Canada
jer@cim.mcgill.ca

Abstract. We consider two problems related to communication between geographically distributed family members. First, we examine the problem of supporting peripheral awareness, in order to improve both emotional well-being and awareness of family activity. This is based on a field study to determine the role and importance of various peripheral cues in different aspects of everyday activities. The results from the study were used to guide the design of our proposed augmented communications environment. Second, we consider the choice of mechanism to facilitate the on-demand transition to foreground communication in such an environment. The design suggests an expansion of Buxton's taxonomy of foreground and background interaction technologies to encompass a third class of *peripheral* communications.

Keywords: Telecommunications, peripheral cues, geographically distributed family.

1 Introduction

The number of families that live apart, either by choice or necessity, has been increasing due to various social circumstances. In the United States, the proportion of nuclear family households (two married parents and a child) dropped from 40% of all households in 1970 to 23% in 2005, while in the same period, the number of single-adult households climbed from 16% to 28% in 2005 [1][2]. Similarly, in Japan, between 1995 and 2005, there was a drop of 2.5% in the percentage of nuclear family households, while the percentage of people living alone increased by a staggering 28.6% [3]. Additional trends, in part due to greater longevity, indicate a growing elderly population [4], often living in isolation from the rest of their families. These dramatic shifts in household composition entail significant changes in the nature of family relationships and, we argue, place increased importance on the role of communications technology for social benefit.

Telephone conversations and videoconferencing provide the means for voice or video communication, but these are often brief in duration, sporadic, and fail to

T. Okadome, T. Yamazaki, and M. Mokhtari (Eds.): ICOST 2007, LNCS 4541, pp. 81–89, 2007.
© Springer-Verlag Berlin Heidelberg 2007

convey much of the rich background or peripheral information we experience about each other when living together. Furthermore, family members, in particular seniors, may be hesitant to initiate contact using telecommunications technologies, even when they wish to see or speak with their loved ones.

This paper describes our initial effort to compensate for these limitations and engender a greater sense of social proximity to distributed family members. The research to date consists of a field study of peripheral communication cues in family relationships and the design phase of an augmented communications environment that facilitates the exchange of certain peripheral information and allows seamless transitions between peripheral and foreground communications.

1.1 Peripheral Communication

In everyday life, people who live together consciously or unconsciously convey, perceive, and share various peripheral information. Examples include the cues of tone of voice, singing in the kitchen or shower, the pace of footsteps, doors being opened or slammed shut, light or music leaking through a door, movement of personal belongings such as bags and keys, and the aroma of coffee brewing or cookies baking. Of course, such cues may have divergent interpretations and significance to different family members and across different families. It is thus necessary to understand how individuals might interpret and use specific cues, perhaps subconsciously, to gain awareness of the mood or physical presence of other family members.

When family members move apart, these cues are no longer shared, which, we believe, diminishes the sense of close contact the family previously enjoyed. Our research project is investigating how technology can help convey these subtle, but important elements of peripheral information to family members or partners living apart for extended periods. Our hypothesis is that the exchange of appropriate peripheral cues will lead to an improved awareness of each others' mental and emotional states, and in turn, reduce the burden on individuals when they wish to initiate communication, thereby leading to improved contact between family members. If implemented correctly, this may improve emotional connectedness, decrease feelings of loneliness and separation, and augment conventional (foreground) communications technology, leading to more productive and fulfilling interactions.

1.2 Previous Literature

Home technologies that aim to assist family members living apart, in particular, seniors, have been investigated by other HCI researchers. Efforts in this area include the use of various digital props, for example, life-size cardboard cutouts of family members [5], family portraits [6], Internet teapot [7], Message Center [4], interactive light table [15] and the installation of an augmented "planter" [8] that senses and conveys physical motion and touch of remote family members. Despite the apparent simplicity of these devices, family members reported powerful emotional affects resulting from their placement in the home.

The major research question relates to the determination of whether more significant cues are sensed and conveyed to remote family members in a meaningful form so as to increase peripheral awareness of the state of loved ones without the technology becoming intrusive or overly demanding of foreground attention.

2 Studying Peripheral Cues Among Family

Field studies of technology use in the home have been conducted for a variety of purposes include natural observational studies of family awareness [9] and information organizing systems in the home [10][11]. Social and emotional factors have also been considered within the eldercare experience of "aging in place" [12]. However, there have been comparatively few studies on the details of background communication among close individuals [8]. In the design of an effective communications environment for close individuals who live apart, we think it is important to understand how various communication cues are used by these individuals while living together.

2.1 Method and Research Settings

As a first step, we conducted a field study consisting of a series of empirical sessions, involving interactive semi-structured interviews, a set of questionnaires, in-situ contextual inquiry sessions, and open-ended discussion. The aims were to understand the participants' current use of communications media, determine important peripheral cues for sensing presence and mood of family members, memory triggers that evoke feelings of missing one another, and verify that our assumptions concerning the use of peripheral communication were valid.

Seven respondents (two male and five female), ranging between 19 and 26 years of age, participated. Table 1 summarizes the profile data of these individuals (indicated by initials) as well as a breakdown of their use of peripheral cues in relating to other family members.   With only one exception, the participants described their relationships with family members as open, relaxed, devoted, and involving frequent communication. The field study took place in Montreal, Canada between July and September of 2006 and involved approximately nineteen hours over nineteen sessions in total. The sessions, spread over several weeks, were conducted on an individual basis to assure participants' privacy and divided into three components. Each such component involved a period of discussion, completion of a questionnaire, and a take-home data gathering exercise.

The aim of the introductory session was to facilitate for participants a better understanding and awareness of their everyday background communications with family members and establish an appropriate rapport. The initial questionnaire (Q1) involved topics of family members' profiles, feelings regarding relationships with family members, and current use of communications media. Prior to the second session, participants were asked to complete a second form (Q2) by listing a number of cues related to everyday background communications with family members at their home. In order to gain specific information regarding the various roles that peripheral communication plays in everyday activities, these were divided into categories of gaining awareness of (a) mood and (b) physical presence of family members, as well as (c) memory triggers that evoke feelings of missing family members.

During the second session, typically one week later, we began with a debriefing of the responses to Q2 and followed this with another questionnaire (Q3) concerning participants' current feeling of loneliness and well-being. This was done to understand their potential motivation of further contacts with family and the context for their peripheral communication cues. Prior to the third session, participants were asked to complete another form (Q4, which was an elaboration of Q2), listing as many peripheral cues as possible. Participants were also asked to draw floor plans of their home and capture video or photographic examples as helpful to illustrate their list of cues. These were expected to provide a basis for understanding the spatial relationships of peripheral cues and assist our design of a prototype augmented environment for peripheral communication.

During the third session, again, typically one week later, we conducted semi-structured in-situ contextual interviews to evaluate the level of emotional responses and their importance, as related to the list of cues obtained from Q2 and Q4.

Table 1. Profiles of respondents and summary of peripheral communication cues

	Gender	Age	Living apart from which family members	Living with	Peripheral communication cues					
					Visual	Audio	Somato sensory	Olfactory	Taste	Other
FR	M	26	brother, parents	n/a	14	14	0	0	0	0
NR	M	23	brother	parents	17	16	0	1	0	0
MK	F	19	parents	n/a	9	16	1	0	0	3
TR	F	22	partner, parents	n/a	4	3	0	1	0	0
IM	F	22	partner	roommate	13	9	2	0	0	4
CY	F	23	partner	n/a	9	13	0	3	0	2
LP	F	23	partner	parents	4	3	0	0	0	0

2.2 Results

Use of Conventional Communications Media. From the comments and discussion with study participants, several shortcomings of conventional (foreground) communica-ions media were noted. In general, emotional characteristics of communications, including one's overall mood and expressions of sarcasm or humor, were felt to be not as easily conveyed through communications media as in person. The two most popular forms of communications media used by our study participants were clearly telephone and email, with the former being preferred, when available, almost exclusively. Nevertheless, one participant commented explicitly on the inadequacy of email for expressing feelings, in particular as she spoke a different mother tongue from her partner, while another noted that misunderstandings may arise from the lack of visual cues available in telephone conversation. Additional issues raised regarding telephone communication concerned the restrictions on engaging in parallel activities, such as washing dishes, due to background noise. The danger of misinterpretation, in particular for jokes or sarcasm, was also raised in regard to instant messaging. As has been described in numerous other studies, social communication relies to a great deal on visual cues to provide context for verbal remarks.

Peripheral Cues. A total of 161 distinct peripheral communication cues were obtained from the responses to our questionnaires Q2 and Q4. These were analyzed according to the primary modality of individual cues and classified into six categories of visual, audio, somatosensory, olfactory, taste and others (Table 1).

The results indicate that audio (46%) and visual (43%) were the two dominant modalities of peripheral communication cues with family. Very few cues were reported for the other modalities of somatosensory (1.9%), olfactory (3.1%) and taste (0%) information. A small number of cues involved the description of an overall experience, typically a non-instantaneous event such as going for a walk, or to a dim sum restaurant. As these involved multiple modalities over a particular interval, or did not involve any particular modality information, these were classified in a separate category of "others" (5.6%).

An analysis of the distribution of cues by category (Table 2) suggests that the two dominant modalities of audio and visual cues may play different roles in peripheral communication. Audio seems to be particularly relevant for gaining awareness of the mood of other family members, whereas visual cues were more strongly related to triggering memories that evoke feelings of missing one another. Both modalities were equally significant in relation to cues of physical presence of other family members.

Table 2. Peripheral cues summarized by category

	state of mind	physical presence	thinking of family	Sum
Visual	12	26	32	70
Audio	33	26	15	74
Somatosensory	0	0	3	3
Olfactory	0	2	3	5
Taste	0	0	0	0
Others	2	0	7	9

3 Peripheral Communications Prototype

Based on the results of our initial field study, we attempted to specify the design of an augmented environment for the transmission of peripheral cues between geographically separated family members.

3.1 Modality Considerations

The preceding analysis of cues suggests the dominance of visual and audio modalities in peripheral communication, with audio particularly important for conveying a sense of feelings or emotions of family members. Since our primary intent is to support a greater awareness of the well being and feelings of distant family members, it therefore makes sense to focus on the extraction of relevant auditory cues from the environment and the manner in which these are delivered to the remote party. An additional benefit of working with the auditory, rather than visual, modality is that rich information may be conveyed without requiring family members to be in a particular room where a display is visible.

Relevant audio cues cited by our study participants included the gait of footsteps, doors being slammed shut, music played, and the sounds of a meal being prepared, typically as indicators of presence of other family members. Similarly, the tone, inflection, and rate of speech were mentioned as strong cues of emotional state, even if individual spoken words cannot be recognized. These observations are significant, as they suggest that audio may afford a high degree of peripheral communication even if reproduced in a low-fidelity manner, provided that the salient characteristics are preserved. At the same time, it is necessary to ensure that our system neither conveys foreground information, such as clearly discernable speech, nor commands explicit attention. This may be achieved by some form of active content filtering, or by mapping audio input to some other modality, for example, an abstract graphic visualization, similar to those provided by music player software. Miyajima et al. provide such a mapping between the input of touching a terminal and the corresponding output sounds produced at the remote location [8]. In contrast, we are motivated to convey cues in their original modality, preserving as much richness as possible in the communication. In either case, it is humans, rather than technology, that are responsible for the interpretation of these cues. However, we believe that such interpretation is severely limited when it cannot benefit from the skills that family members develop through years of experiencing cues in their original form.

3.2 Peripheral Audio

Our design aims to convey the maximum information content without crossing into the domain of foreground communication. To do so, we propose to transmit muffled audio between the two sites, simulating the perceptual effect of hearing the sounds made by family members several rooms away; in other words, virtually extending the house in the manner of a geographically distributed soundscape. This may be accomplished easily by damping the microphones in order to mimic the acoustic effects one would experience as sound travels through walls. This would offer the benefit of relegating what may be a complex computational filtering task to natural physics. To preserve meaningful semantics, we believe that sound source location and directionality are important, in the sense that sounds from the kitchen may not be perceived in the same context if they are reproduced in a different room at the remote location. Thus, use of multiple microphones and speakers is required, with an attempt to match the sources and sinks to socially equivalent locations at the two sites. Automated gain control might be utilized to balance reproduced audio cues with the local ambient volume level. In addition, manual gain control may prove important to users, at least for acceptability of the technology as deployed in a home environment. Similarly, privacy concerns are reduced because muffled speech remains unintelligible.

4 Transitions Between Periphery and Foreground (and Back)

Buxton [13] proposed a taxonomy for classifying communications technologies, divided into human-human/human-computer along one axis, and foreground/background along the other. His definition of *background,* like ours, relates directly to activities that take place in the periphery of human attention. While Buxton's model includes smart-house and the "Portholes" system [14] as examples of background human-computer and human-human interaction, respectively, these nevertheless demand a certain level of focused user attention to issue commands or queries, but also, in the case of Portholes, at least, to interpret current state. Moreover, exploiting the (computer-mediated) human-human technologies, even in a purely background context, rely on a centralized element of information technology, typically a computer display, which assumes a relatively stationary user. This may be acceptable in an office scenario, but is unlikely typical of a home environment, the primary domain of our interest. We believe that truly *background* technology must be able to convey meaningful state information without conscious effort on the part of its users *and* allow accessibility to this information from a wide variety of locations within one's environment. Portholes does not necessarily meet either of these definitions, while the Ambient Scribbles of the RemoteHome [15] violates the latter definition as it can only be viewed by users facing the correct direction. To distinguish between these cases, we propose the following refinement of Buxton's taxonomy, in which technologies that operate *entirely* in the user's periphery of attention are deemed *peripheral* (Figure 1).[1]

Buxton's model not only provides a taxonomy but describes the transitions between states in response to various events. Keeping with this approach, our system, placed in the newly added third column, supports transitions as follows. A user who is currently receiving peripheral audio from a remote family member may request initiation of foreground human-human communication with that individual by entering the specific room from which the sounds are currently heard, and performing an appropriate gesture or speaking a passphrase. This information is received by the distributed technology (peripheral human-computer interaction). The gesture might take the form of an arm wave, as if trying to capture someone's visual attention, while the passphrase could be an explicit "Hi, Dave!" Clearly, these options would require additional system components such as image processing or speech recognition, and would need to be robust to potential false positive recognition. A simpler alternative would be to equip each room with a tangible interface, for example, a mock-up of a telephone, which, when picked up, or perhaps, simply touched (background human-computer interaction), would serve as the indication of user intent to establish a foreground connection with the other family member.

At the remote end, either an iconic sound (e.g. "ringing") or visual cue that reminds a receiver of the caller (background human-computer interaction) could be generated to indicate that someone wishes to speak with them. However, it is worth reflecting on Weiser and Seely Brown's definition of calm technology as that which

---

[1] Such a definition might equally well be considered as pervasive or ubiquitous computing, but given the proliferation of technologies under this label that fail to meet our requirements, we prefer to avoid potential confusion.

	Foreground	Background	Peripheral
Human-Human	Face-to-face conversation, telephone, video conference, email	Communicating background information using a centralized interaction device (e.g. Portholes, Digital Family Portrait, Family Planter)	Communicating background information using peripheral interactions anywhere within an environment (e.g. pervasive sounds of remote family members)
Human-Computer	GUIs	Smart-house technology, or indicating intention by manipulation (e.g. by lifting a telephone handset)	Communicating intentions using natural indicators (e.g. intent to establish foreground connection to remote family member)

Fig. 1. Revised model of communications technologies (based on Buxton [13])

"engages both the center and the periphery of our attention, and in fact moves back and forth between the two" [16]. Ideally, we would like the *calling cue* to fit this role, rather than being a disruptive signal, although this entails the risk that the caller may be inadvertently ignored. One option would be to convey the unfiltered sound of the caller's voice, at an appropriately discrete level, as they speak the passphrase, thereby signaling the caller's intent. With sufficient audio resources, this might be enhanced by spatializing the sound so that the caller appears to have moved closer to the intended receiver. The actual connection (foreground human-human interaction) would only be established if the called party accepts the request, as indicated by a corresponding gesture, action, or utterance. At that point, the communications technology would temporarily stop filtering the transmission of audio between the two sites, at least for the two rooms currently occupied by the family members at each end, and the communication link becomes a conventional speakerphone.[2] It may be desirable for the system to permit users to move around their respective environments, while maintaining the foreground communication they have already established, although privacy considerations must be kept in mind when other family members are also present.

Similarly, a transition from foreground back to peripheral communication could be effected by a wave or utterance "goodbye," or equally, the action of returning the mock telephone handset to its base. In either case, the transition between states is easily established, making use of similar cues to those employed in the everyday world.

5 Concluding Remarks

With the overall design now completed, we are beginning the effort of translating these ideas into a practical implementation that will be deployed in the homes of some of our initial study participants. These individuals will remain actively involved in the development process by providing continued feedback as the design evolves, Of

---

[2] Issues of half-duplex communication or echo-cancellation processing must be considered in order to avoid the problems of feedback.

particular interest, we wish to experiment with the various options for initiating transitions between states, as described in the previous section, in order to evaluate their ease- and frequency of use. We expect to conduct multiple evaluations of family interaction, before, during, and after deployment, in order to assess the effectiveness of our prototype. This feedback will, in turn, drive successive iterations of the technology. We hope that this effort will prove beneficial in helping distributed family members maintain awareness of each other's state and emotional well-being in a manner that goes beyond the capabilities of current foreground communications technology.

Acknowledgments. We thank our study participants who generously volunteered their time to provide us with detailed insight into the role of various peripheral cues in family communication.

References

1. US Census Bureau datasets, available from http://factfinder.census.gov
2. http://www.answers.com/topic/family
3. Population Census of Japan, Statistics Bureau, Ministry of Internal Affairs and Communications, Japan, translated from
   http://www.stat.go.jp/data/kokusei/2005/kihon1/00/04.htm
4. Wiley, J., Sung, J., Abowd, G.: The Message Center: Enhancing Elder Communication. In: Proc. ACM CHI'06, Work-in-Progress (2006)
5. The Associated Press, Wire Report, Life-sized likenesses of Guard members ease separation pains (August 29, 2006)
6. Mynatt, E.D., Rowan, J., Jacobs, A., Craighill, S.: Digital Family Portraits: Supporting Peace of Mind for Extended Family Members. In: Proc. ACM CHI '01 (2001)
7. Internet Tea Kettle http://www.mimamori.net/
8. Miyajima, A., Itoh, Y., Itoh, M., Watanabe, T.: "Tsunagari-kan" Communication: Design of a New Telecommunication Environment and a Field Test with Family Members Living Apart. Intl. J. of Human-Computer Interaction 19(22), 253–276 (2005)
9. Hutchinson, H., Mackay, W., Westerlund, B., Bederson, B.B., Druin, A., Plaisant, C., Beaudouin-Lafon, M., Conversy, S., Evans, H., Hansen, H., Roussel, N., Eiderback, B. Technology Probes: Inspiring Design for and with Families. In: Proc. ACM CHI '03 (2003)
10. Taylor, A. S., Swan, L.: Artful Systems in the Home. In: Proc. ACM CHI '05 (2005)
11. Nomura, S., Tamura, H., Hollan, J.: Information Management Centers in Everyday Home Life. In: Proc. HCI International 2005, LEA (2005)
12. Hirsch, T., Forlizzi, J., Hyder, E., Goetz, J., Stroback, J., Kurtz, C.: The ELDer Project: Social and Emotional Factors in the Design of Eldercare Technologies. In: Proc. Conf. CUU'00 (2000)
13. Buxton, W.: Integrating the Periphery and Context: A New Model of Telematics. In: Proceedings of Graphics Interface '95, pp. 239–246 (1995)
14. Dourish, P., Bly, S.: Portholes: Supporting Awareness in a Distributed Work Group. In: Proc. ACM CHI'92, pp. 541–547 (1992)
15. http://www.remotehome.org/
16. Weiser, M., Seely Brown, J.: Designing Calm Technology, Xerox PARC (December 21, 1995) http://www.ubiq.com/hypertext/weiser/calmtech/calmtech.htm

An Authentication Architecture Dedicated to Dependent People in Smart Environments

Mhamed Abdallah, Cecilia Fred, and Arab Farah

HANDICOM Labs - INT Evry
{abdallah.mhamed,fred.cecilia,farah.arab}@int-evry.fr

Abstract. Nowadays the concept of "pervasive computing" is fully deployed in smart environments to bring more comfort and to allow an easy access to services for users. Due to this fact it is essential to reconsider some security service delivery in smart environments. The authentication service is tightly related to the user and must imperatively take into account his capacities, his preferences and his environment, to be efficient for dependent people. In this paper we propose a new architecture able to provide an intelligent authentication based on modelisation parameters of both the environment and the user profile which are used to build an authentication strategy.

Keywords: Pervasive computing, smart home, user profile, dependent people, security, authentication, biometrics.

1 Introduction

The concept of *ubiquitous computing* more generally known as *pervasive computing* represent a new vision of the access to information and interaction between the man and his environment. The "traditional" environments are becoming more and more "intelligent", due to the small dimensions and high performances of tiny processors which can be easily integrated in everyday life objects and can spontaneously communicate with each others. Some of recent main research projects dealing with security in pervasive computing which we have investigated are the "Oxygen" project of MIT[1] [1][2] organised into two subprojects which are "Mobile network device" and "Intelligent Room", the "Aware Home Rechearch Initiative" project[3] of the technological institute of Georgia, the "GAIA" project of the university of Illinois [4][5], and the "Easy Living" projet [6] of Microsoft. Among them, only "GAIA" project gives specifications about its authentication architecture which brings some interesting concepts but does not take into account the user profile. This paper lies within the scope of the "Smart Home" project devoted to the creation of generic tools for technical assistance to improve the life of dependent people (old and disabled) within intelligent environments. Due to the heterogeneity and complexity of existing authentication devices/protocols, it is necessary to reconsider security requirements of dependent people in such environments [7]. To meet the specific needs for dependent

---

[1] Massachussets institute of technology.

T. Okadome, T. Yamazaki, and M. Mokhtari (Eds.): ICOST 2007, LNCS 4541, pp. 90–98, 2007.
© Springer-Verlag Berlin Heidelberg 2007

people, it is essential to have a more dynamic vision which require new security models able to adapt to the profile, the context and the environment of these users. Consequently, in our approach we made an exploratory study to derive the required parameters specific to dependent people and which have to be considered for the design of an authentication architecture both "intelligent" and "generic".

The remainder of this paper is organized into four sections. In section 2, we will show an exploratory study about the usability of authentication devices by disabled people. Section 3 is dedicated to the description of our proposed architecture model while section 4 detail its implementation. Finally, section 5 will give the limits of this work and discuss about its evolution.

2 Authentication Device Usability by Disabled People

Several authors [8] consider the utility of technical devices as not sufficient to justify their use even, if they meet the user's needs. They have shown that accessibility and usability defects can also compromise their use. Therefore, the aim of our study was concentrated on using authentication devices by disabled people to evaluate difficulties which can likely be encountered regarding to their impairments and their incapacities.

We have considered eight means/objects allowing the user authentication: the keyboard (seizing password), the smart card , the fingerprint, the face, the hand, the voice, the retina and the iris.

From our study we have extracted some relevant data[2] which allows us to make a first evaluation of usability authentication devices by disabled people (fig 1). The use of each authentication device was investigated relatively to 36 types of motor disabilities. Figures 1, show the results associated to the number of times these devices were considered as being "avoiding", "possible" or "desirable".

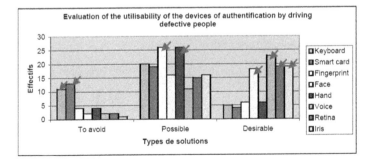

Fig. 1. Authentication device usability

In order to determine the links between the authentication devices and their appreciation level by people with motor disability, we have used Cramer's

---

[2] Data analyzed with the software DS3-WIN.

contingency coefficient.[3] The results show that the usability of the authentication devices depends on the parts of the body which are requested. Generally,
the usability of biometric devices using the face, the voice, the retina or the iris
seems to be less constraining for disabled people Nevertheless, we observe a more
important rejection of devices such as the keyboard or the smart card, since they
require further mobilization of the body parts. In this case, the defects of device
usability are due to the lack of compatibility between their design features and
the user's capacities. In addition, we can observe that in more than half of the
cases (51,74%), the use of some authentication devices (keyboard, smart card,
fingerprint, hand) is considered as possible. Indeed, the usability of some devices
is strongly related to the degree of disability.

This exploratory study highlighted the need to adapt the biometric systems
to the users and the importance to get reliable and effective devices, able to
adapt to user's variability. Ergonomic evaluations in real situation with disabled
users are necessary for taking account the real constraints and requirements of
using authentication devices. They should be performed in order to refine this
study by considering some other parameters related to the context (noisy or sink
environments), to the incapacity degree of involved people, to their age, etc...
According to this primary study, it is necessary consider account the user profile
in the design of our authentication architecture.

3   Proposal for a New Authentication Architecture

3.1   Model Description

As illustrated in fig 2, the model that we propose is based on an architecture
which requires an interaction between the three entities described below: the
user device, the smart card and the authentication service.

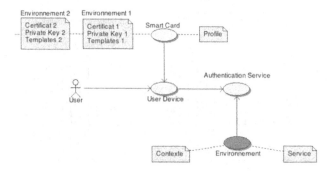

Fig. 2. Model entities

---

[3] The Cramer's contingency coefficient calculated is 0,1356 (Reference marks: $0,04 < V^2 < 0,16$).

- **The user device:** which can be a Tablet PC, a PDA or even a mobile phone. It allows the interaction between the system and the user who will be able to ask for any service available in the environment and also receive the required instructions for his authentication.
- **The Smart Card:** which contains all the parameters allowing a user to be authenticated. Moreover, all cryptographic calculations will be made on the processor of the card. Assuming that the user can be called to attend several intelligent environments (place of work, house, hospital, etc), and that these environments can use different devices and algorithms (particularly for biometric systems), the card will have a manager of environment. This manager supervise the various environments of the user; each environment is related to a set of parameters allowing user authentication. The manager has also to store automatically these parameters in memory when needed.
- **The Authentication server:** after a required number of checking, the authentication server delivers a ticket to the user, granting him the access to some services within the "smart home". These tickets have an expiry date and can be provided only by the authentication server. It will set up a strategy allowing the checking process according to the user profile, the context, as well as specific needs of the required service.

3.2   Initialization Process of an Authentication Session

This architecture takes into account the user profile when setting up the authentication strategy. From the user profile, one will be able to know the devices which can be usable for the authentication. Information coming from the context makes it possible to evaluate the relevance of using these devices by allocating a "confidence value" to each device. Compared to the "confidence value" of GAIA project, the one we have used for our model is dynamic and depends on the context. For example, if the user profile allows an authentication by using voice recognition device, using context parameters we will be able to determine if it is relevant or not to use this device according to the ambient noise. Each service has a profile, to which we associate a "confidence threshold ". This threshold represents the level of certainty to be reached about the user identity for getting access to the required service. By dealing with various authentication devices, the system must develop an authentication strategy which help to reach the required confidence threshold, by cumulating confidence values of selected devices. Indeed progressively , as long as the user stay within the same environment, the smart card increment a parameter called "user value" representing the greatest threshold of confidence reached for that user during the current session.[4]

---

[4] One session begins at the first request of ticket and finishes when the user leaves the environment or if he remains inactive during a certain time.

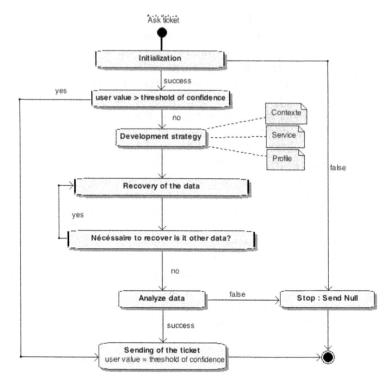

Fig. 3. Authentication session

4   Implementation

4.1   Class Description

The prototype implementing our architecture was developed in java langage. We
used RMI[5] for the interactions between the authentication service and the user.
This prototype is built using the four main classes shown in (fig 4).

The **Smart_Env class** modelise the set of parameters which allow the user
authentication in a given environment (house, work place, etc). It has the six
following elements:

- **The certificate** provided by the authority supervising the environment,
- **The Private key** associated to the public key included on the certificate,
- **The user profile** represent the set of user characteristics (capacities,
  preferences),
- **The templates** which are used as a reference for biometric authentication,
- **The hash of the password** chosen by the user,
- **The User value** which is used to save the confidence value.

---

[5] Remote Method Invocation.

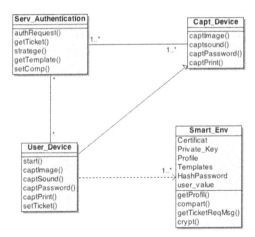

Fig. 4. Class Diagram

The first four elements are used during the first authentication phase which will be described in next sub-section. The function getTicketReqMsg() generate the request message for a ticket.

The function "crypt()" is called when using a biometric modality. It allows sending its captures to the authentication service for manufacturing the template to be checked. Indeed, in order to provide an architecture as generic as possible, the manufacture of the templates is dealt by the authentication service. The function "compart()" makes the comparison between the data entered by the user and the data stored on the smart card (template, hash of the password). It returns an encrypted message with the result of this comparison. The function "getProfil()" recover the user profile when requested by the service.

The Capt_Device class manages the capture functions and can be allocated to any available authentication device.

The User_Device class is a sub-class of Capt_Device class. It integrates all the functions for the capture, for starting the authentication, for interaction with the smart card and for ticket granting.

The Serv_Auth class class corresponds to the authentication service.

This class is modelized using three layers. The low layer allows the interactions with the environment and the user for the acquisition of information and the capture recovery. The intemediate layer may be similar to a tool-layer, including the "strategist" entity which receive data, create an authentication strategy, and select the best algorithms for the template creation. The top layer represents the authority of decision-making.

This class is composed of 5 main functions: authRequest(), getTicket(), stratege(), getTemplate() and setComp(); We will detail the "authRequest()" and "getTicket()" functions in the following section. The "stratege()" function is dealing with the strategy generation based on the user profile.

The function "getTemplate()" allows the call of the template's manufacturing process. Lastly, the function "setComp()" recovers the encrypted messages sent by the user in order to inform the service about the results of the comparison.

4.2   User Authentication Process

Obtaining a ticket for using a service occurs in the 3 phases illustrated in figure 5.

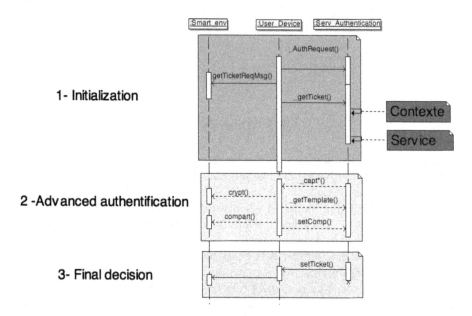

Fig. 5. User Authentication Process

Phase 1 - Initialization: The aim of this phase is to certify that one can trust the smart card before allowing key exhanges in a secured way. Its implementation is based on an extension of Kerberos protocol[9] called PKDAKerberos[10]. PKDAKerberos keep the concept of ticket delivery at the end of the authentication process, while extending it for a distributed architecture by using asymmetrical keys. This protocol is triggered by the user and occurs in five exchanges. We kept the first three exchanges of the PKDAKerberos protocol, while making some modification during the third exchange by adding :

- A "user value" which is used to get the highest confidence threshold reached during a session, to determine whether it is necessary to make an advanced authentication or not.
- A second symmetrical key used to encrypt the messages exchanged during the second phase of advanced authentication .

Phase 2 - Advanced authentication: During this phase the service involves an intelligent entity called the "strategist". Its main tasks are first to recover the relevant parameters of the user profile, to determine the devices which can be used by the user (beta strategy). Once this "beta strategy" is built, the strategist analyse the profile parameters and the environment related data to refine and improve the strategy. These data were taken from the results of work in context awareness given in [11]. Obviously the quantity and the precision of information given by the user profile will allow deeper analysis and better results closer to reality. In addition further investigation based on the exploratory study of Section 2 would be essential for the training process. Thereafter the authentication service will call upon the capture functions (captImage(), captSound, captPassword() captPrint()) selected by this strategy. It would be extremely interesting to make this choice regarding to both the performances and the availability of devices involved in the service. The last stage is the manufacturing of the templates (if biometrics is used) followed by various comparisons and the transmission of a scoring report.

Phase 3 - Final decision: This phase is equivalent to the 4th exchange of the PKDAKerberos protocol. The user will then be able to use the granted ticket for the access to the required service as in the 5th exchange of PKDAKerberos.

5   Conclusion and Perspectives

To provide authentication of dependent people, we have proposed an new dynamic and extensible architecture to allow a flexible and smart authentication process. Using specific parameters modelizing the user profile and environment, the authentication server will set up an authentication strategy accordingly to the user context. Based on this strategy, the user can be authenticated by a multimodal system combining different means (password, smart card, biometry). In order to cover various smart environments, this architecture can be implemented either in a centralized or distributed environment. The current prototype allows an authentication by using the password and has two modules for voice and image capture. Future evolution of the project will integrate the smart card module and further biometric modules. The implementation of the strategist would require an in-depth study about the characteristics of involved population and available devices.

References

1. Coen, M.: Design principles for intelligent environments. In: Proceedings of the fifteenth National Conference on Artificial Intelligence (AAAAI98), Madison, WI, pp. 547–554 (1998)
2. Burnside, M., Clarke, D., Mills, T., Maywah, A., Devadas, S., Rivest, R.: Proxy based security protocols in networked mobile device. In: Proceedings of ACM Symposium on Applied Computing (SAC 2002), Madrid, Spain (March 2002)
3. Matthew, M.C.: Generalized role based access control for securing future applications. In: Proceedings of the 23rd National Information Systems Security Conference, Baltimore, MD (October 2000)

4. Gill, B., Viswanathan, P., Campbell, R.: Security architecture in GAIA, Technical report, UIUCDCS-R-2001-2215 UILU-ENG-2001-1720, University of Illinois at Urbana, Champaign, IL, USA (2001)
5. Al-Muhtadi, J.F.: An Intelligent authentication infrastructure for ubiquitous computing environments, PhD thesis, University of Illinois, Urbana-Champaign, IL, USA (2005)
6. Hedberg, S.R.: After desktop computing: a progress report on smart environments research. IEEE Intelligent Systems and Their Applications Magazine 15(5), 7–9 (September-October 2000)
7. Mhamed, A., Mokhtari, M.: Providing a new Authentication Tool for Dependant people in Pervasive Environments. In: Proc. of the IEEE International Conference on Information & Communication Technologies (ICTTA04), Damascus, Syria (April 2006)
8. Specht, M., Burkhardt, J.M., De La GarzaC.: De lactivit des ans confronts aux nouvelles technologies, Retraite et Socit, Technologie et Vieillissement, vol. 26/2, pp. 21–38
9. Neuman, B.C., Theodore, T.: Kerberos: An authentication Service for Computer Networks. IEEE Communications Magazine 32(9), 33–38 (September 1994)
10. Sirbu, M.A., Chung-I Chuang, J.: Distributed Authentication in Kerberos Using Public Key Cryptography, Internet Society 1997, Symposium on Network and Distributed System Security (February 1997)
11. Feki, M.A., Mokhtari, M.: Context awareness for pervasive assistive environment, in IDEA publisher (ed.) Ismail Khalil Ibrahim, Handbook of Research on Mobile Multimedia. Idea Group Publisher (May 22, 2006) ISBN: 1-59140

Secure Spaces: Protecting Freedom of Information Access in Public Places

Shun Hattori and Katsumi Tanaka

Department of Social Informatics, Graduate School of Informatics, Kyoto University
Yoshida-Honmachi, Sakyo, Kyoto 606-8501, Japan
Tel.: +81-75-753-5385, Fax: +81-75-753-4957
{hattori,tanaka}@dl.kuis.kyoto-u.ac.jp

Abstract. In public places, information can be accessed by its unauthorized users, while we are sometimes forced to access our unwanted information unexpectedly. We have introduced the concept of Secure Spaces, physical environments where any information is always protected from its unauthorized users, and have proposed a content-based entry control model and an architecture for Secure Spaces which protect their contents' freedom of information delivery but do not protect their visitors' freedom of information access. This paper aims at building truly Secure Spaces which protect both freedoms and enhances our proposed model.

1   Introduction

Today, in public spaces which are shared by the general public or shared spaces (e.g., meeting room at office or living room at home) which are shared by specific multiple persons (e.g., co-workers or family members), information resources can be usually accessed by multiple users, not only their authorized users but also their unauthorized users and thus it is not always enough for their security policies, while visitors are sometimes forced to involuntarily access various information resources, not only their wanted information resources but also their unwanted information resources and thus it would be not good for their health.

In our previous work [1,2], we have introduced the notion of Secure Spaces as physical environments where any information resource is always protected from its unauthorized users' eyes and ears, and have proposed a model and an architecture for space entry control based on its dynamically changing visitors and contents such as physical information resources and virtual information resources via embedded output devices, unlike the conventional context-unaware access controls. The outdated Secure Spaces, however, do not protect their visitors' freedom of information access. In this paper, we aim at building really Secure Spaces which protect both freedoms, and revise our proposed model.

The rest of this paper is organized as follows. First, Section 2 describes requirements for building Secure Spaces. Next, Section 3 gives an overview of our content-based entry control, Section 4 formalizes a model, and Section 5 introduces an architecture. Subsequently, Section 6 presents some related works. Finally, we conclude this paper in Section 7.

T. Okadome, T. Yamazaki, and M. Mokhtari (Eds.): ICOST 2007, LNCS 4541, pp. 99–109, 2007.

2   Requirements for Secure Spaces

We redefine Secure Spaces as physical environments which protect not only their contents' freedom of information delivery but also their visitors' freedom of information access, and then discuss the requirements for building Secure Spaces.

We have had not only immobile and personal computing environments but also mobile and ubiquitous computing environments, and thus we have become able to access computational information not only in private spaces but also in public spaces or shared spaces such as meeting rooms or living rooms.

In private spaces where a computer is exclusively used by a single occupant, we can assume that information outputted via the computer is received by only the person who operates the computer. Therefore, information access control systems have only to take into account the single user directly operating the computer and requesting to access a virtual information resource in making an authorization decision on whether the access request should be granted or denied.

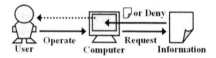

Fig. 1. Single information receiver in private spaces

Meanwhile, in public or shared spaces where a computer can be shared by multiple visitors, we can assume that information outputted via the computer is received by not only the single user directly operating the computer and requesting to access it but also the other users nearby surrounding the computer. Therefore, it is not enough for information security only to check on whether a user requesting to access a virtual information resource via a computer is in the set of its authorized users, because there might exist its unauthorized users in the area surrounding the computer and they might be able to access it.

We would have the following approaches for securing public/shared spaces. By using any approach, Secure Spaces have to be always aware of their real-world contexts such as visitors, physical information resources and virtual information

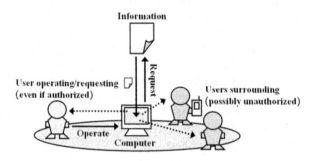

Fig. 2. Multiple information receivers in public/shared spaces

resources outputted via embedded computers. Figure 3 shows each approach for a case that a space where Alice can access Info1 and Info2 outputted via its embedded computers is being entered by Ichiro who don't want to access Info2.

(a) **Logically Access Control:**
  revokes logically problematic information which will be possibly accessed by its unauthorized users or those who don't want to access it if nothing is done. In Figure 3, Ichiro can be protected from his unwanted Info2 if it is revoked. However, Alice becomes unable to access it either even if it is wanted by her.

(b) **Context-aware Media Conversion:**
  presents problematic information in a medium (e.g., languages, audio/visual) where it is perceptible or understandable for any user who don't want to access it and who will possibly access it if nothing is done. In Figure 3 where Ichiro can understand Japanese but not English, Ichiro can be protected from his unwanted Info2 (Alice might be kept accessible to any) if it is presented in not Japanese but English. However, Secure Spaces have to know any user's profile on whether or not s/he can understand information in each medium.

(c) **Physically Entry Control:**
  shields physically problematic information from its unauthorized users and those who don't want to access it by means of opaque and/or soundproof partitions. In Figure 3, Ichiro can be protected from his unwanted Info2 and Alice can be kept accessible to it if Alice and her wanted Info2 are physically isolated from Ichiro. However, Alice might become unable to access Info1.

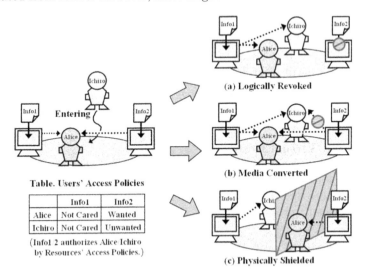

Table. Users' Access Policies

	Info1	Info2
Alice	Not Cared	Wanted
Ichiro	Not Cared	Unwanted

(Info1 2 authorizes Alice Ichiro by Resources' Access Policies.)

(a) Logically Revoked

(b) Media Converted

(c) Physically Shielded

Fig. 3. Approaches for securing public/shared spaces

3 Overview of Content-Based Space Entry Control

This section gives an overview of our content-based entry control for building Secure Spaces which any visitor or information resource is always protected from its unwanted information resource or its unauthorized visitor according to each other's access policy. Our strategy for the requirements is to utilize physical environments (e.g., slightly smart rooms) electrically lockable and physically isolated by opaque and soundproof walls. In addition, we assume that any user inside a Secure Space can access any information resource inside the Secure Space but s/he can never access any information resource outside the Secure Space.

3.1   Entry Control over Physical Resources

When a physical resource (e.g., a hardcopy of sensitive information) requests to enter a Secure Space, the Secure Space will enforce entry control over the physical resource based on the following strategy:

– **Unwanted Physical Resource for Visitors**
   If the Secure Space is containing a visitor who does not want to access the physical resource or who the physical resource does not want to be accessed by, the Secure Space has to deny the physical resource to enter there (possibly after persuading the visitor to relax his/her access policies or to exit there)

Fig. 4. Space entry control over physical resource for users

3.2   Entry Control over Virtual Resources

When a virtual resource (e.g., a sensitive information on computer) requests to be outputted via a device embedded in a Secure Space, the Secure Space will enforce entry control over the virtual resource based on the following strategy:

– **Unwanted Virtual Resource for Visitors**
   If the Secure Space is containing a visitor who does not want to access the virtual resource or who the virtual resource does not want to be accessed by, the Secure Space has to deny the virtual resource to be outputted there (after persuading the visitor to relax his/her access policies or to exit there).

3.3   Entry Control over Users

When a user requests to enter a Secure Space, the Secure Space will enforce entry control over the user based on the following case-dependent strategies:

– **Unauthorized User for Virtual Contents**
   If the Secure Space is containing a virtual resource via an embedded output device which does not want the user to access itself or which the user does not want to access, the Secure Space has two approaches of denying the user

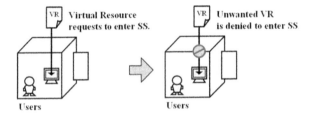

Fig. 5. Space entry control over virtual resource for users

to enter there (after persuading the user to relax his/her access policies) or granting the user after revoking the output session of the virtual resource.
– **Unauthorized User for Physical Contents**
  If the Secure Space is containing a physical resource or a visitor which does not want the user to access itself or which the user does not want to access, the Secure Space has only one approach of denying the user to enter there.

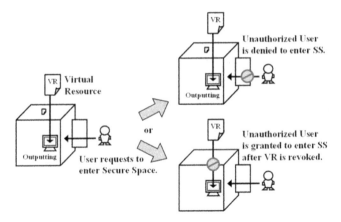

Fig. 6. Space entry control over user for virtual contents

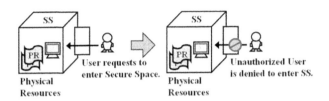

Fig. 7. Space entry control over user for physical contents

4   Formalized Model

This section formalizes our unified model for space entry control based on its dynamically changing visitors and contents such as physical resources and virtual resources via embedded output devices, by listing the entities, functions, access policies, entry requests and authorization algorithms in the model.

Definition 1. Model Entities

Our entry control model for building Secure Spaces has the following four kinds of entities whose access policies should be protected.

- **Secure Spaces:** are physically isolated environments (e.g., a closed room by opaque and/or soundproof walls with electrically lockable doors) which want their Users' and Physical/Virtual Resources' access policies to be protected on our assumption that any content can be accessed by any visitor. The universal set of Secure Spaces is denoted by S.
- **Users:** are physical entities who enter a Secure Space and can access any Resource inside there on our assumption and who can describe their access policies as freedom of information access. The universal set of Users or their Access Policies is denoted by U or UAP respectively.
- **Resources:** are physical/virtual entities which enter a Secure Space and can be accessed by any User inside there and which can describe their access policies as freedom of information delivery. The universal set of Resources or their Access Policies is denoted by R or RAP respectively, where $R = PR \cup VR$ and $RAP = PRAP \cup VRAP$.
  - **Physical Resources:** are physical entities such as a hardcopy of sensitive information. The universal set of Physical Resources or their Access Policies is denoted by PR or $PRAP$ respectively.
  - **Virtual Resources:** are virtual entities such as a sensitive information on the Internet. The universal set of Virtual Resources or their Access Policies is denoted by VR or $VRAP$ respectively.

Definition 2: Model Functions

Our model uses the following functions to keep up on the set of entities inside each Secure Space at a time and evaluate the weight of its contents.

- $cu: S \times T \to 2^U$, is a function mapping each Secure Space s_i at a Time t, to the set of Contained Users inside there $cu(s_i, t)$.
- $cpr: S \times T \to 2^R$, is a function mapping each Secure Space s_i at a Time t, to the set of Contained Physical Resources inside there $cpr(s_i, t)$.
- $cvr: S \times T \to 2^R$, is a function mapping each Secure Space s_i at a Time t, to the set of Contained Virutal Resources inside there $cvr(s_i, t)$.
- $authU: UER \to \{grant, deny\}$, is a function mapping each User's Entry Request uer_i, to the entry decision $authU(uer_i)$.
- $authR: RER \to \{grant, deny\}$, is a function mapping each Resource's Entry Request rer_i, to the entry decision $authR(rer_i)$.
- $w: S \times T \times 2^U \times 2^{PR} \times 2^{VR} \to \mathcal{R}$, is a function mapping a set of Users us, Physical Resources prs and Virtual Resources vrs in each Secure Space s_i at a Time t, to its evaluated Weight $w(s_i, t, us, prs, vrs)$ in some way. In this paper, the weight is defined as the summation of each positive

weight $w(s, t, u, r)$ evaluating that a User u can access a Resource r in a Secure Space s at a Time t (but could be defined in another way),

$$w(s, t, us, prs, vrs) = \sum_{u \in us, r \in prs \cup vrs} w(s, t, u, r),$$

where $s \in S$, $t \in T$, $us \in 2^U$, $prs \in 2^{PR}$ and $vrs \in 2^{VR}$.

Definition 3: Access Policies

Our model stores and protects the following access policies for each entity.

- **User's Access Policy:** is an access policy described by a user in order for him/her to inform Secure Spaces whether s/he wants to access a resource in a context, as a 5-tuple of a User, a Resource, a Secure Space, a Time and a set of surrounding entities as contextual conditions,

$$\text{UAP} \subseteq U \times R \times S \times T \times 2^C,$$

where T stands for the universal set of Times and $C = U$.

$(u, r, s, t, cs) \in \text{UAP}$ where $u \in U$, $r \in R$, $s \in S$, $t \in T$ and $cs \in 2^C$, states that the User u permits accessing the Resource r at the Time t in the Secure Space s which contains the set of entities cs.

- **Resource's Access Policy:** is an access policy described by a resource in order for it to inform Secure Spaces whether it wants to be accessed by a user in a context, as a 5-tuple of a Resource, a User, a Secure Space, a Time and a set of surrounding entities as contextual conditions,

$$\text{RAP} \subseteq R \times U \times S \times T \times 2^C.$$

$(r, u, s, t, cs) \in \text{RAP}$ where $r \in R$, $u \in U$, $s \in S$, $t \in T$ and $cs \in 2^C$, states that the Resource r permits being accessed by the User u at the Time t in the Secure Space s which contains the set of entities cs.

Definition 4: Entry Requests

Our model grants or denies the following entry requests for each entity.

- **User's Entry Request:** is an entry request defined as a 2-tuple of a User and a Secure Space which s/he is requesting to enter,

$$\text{UER} \subseteq U \times S \times T.$$

$(u, s, t) \in \text{UER}$ where $u \in U$, $s \in S$ and $t \in T$, states that the User u requests to enter the Secure Space s at the (current) time t.

- **Resource's Entry Request:** is an entry request defined as a 2-tuple of a Resource and a Secure Space which it is requesting to enter,

$$\text{RER} \subseteq R \times S \times T.$$

$(r, s, t) \in \text{RER}$ where $r \in R$, $s \in S$ and $t \in T$, states that the Resource r requests to enter the Secure Space s at the (current) time t.

Algorithm 1: Authorization for Users

A user's entry request that a User u requests to enter a Secure Space s at a Time t is granted, if and only if any resource inside there permits being accessed by the User and s/he permits accessing any resource inside there in the context or if the weight $w(s, t, u)$ in case of granting him/her to enter there after revoking any virtual resource inside there which denies being accessed by him/her or s/he denies accessing in the context is higher than the weight $w(s, t)$ in case of denying him/her to enter there.

$$\forall u \in U, \forall s \in S, \forall t \in T, \ \text{authU}(u, s, t) = \text{grant}$$
$$\Leftrightarrow (\text{apr}(s, u) = \text{cpr}(s)) \wedge \{(\text{avr}(s, u) = \text{cvr}(s)) \vee (w(s, u) \geq w(s))\}$$

where $\text{au}(s, t, u)$, $\text{apr}(s, t, u)$ or $\text{avr}(s, t, u)$ is the Assumptive set of Users, Physical Resources or Virtual Resources inside the Secure Space s after forcing all physical resources which don't permit being accessed by the User u or s/he doesn't permit accessing to exit there, revoking all virtual resources which don't permit being accessed by him/her or s/he doesn't permit accessing in order to keep secure and granting him/her to enter there, respectively.

$$\text{au}(s, t, u) := \text{cu}(s, t) \cup \{u\}$$
$$\text{apr}(s, t, u) := \{pr_i \in \text{cpr}(s, t) | (pr_i, u, s, t, \text{au}(s, t, u)) \in \text{PRAP}$$
$$\text{and } (u, pr_i, s, t, \text{au}(s, t, u)) \in \text{UAP}\}$$
$$\text{avr}(s, t, u) := \{vr_j \in \text{cvr}(s, t) | (vr_j, u, s, t, \text{au}(s, t, u)) \in \text{VRAP}$$
$$\text{and } (u, vr_j, s, t, \text{au}(s, t, u)) \in \text{UAP}\}$$
$$w(s, t) := w(s, \text{cu}(s, t), \text{cpr}(s, t), \text{cvr}(s, t))$$
$$w(s, t, u) := w(s, \text{au}(s, t, u), \text{apr}(s, t, u), \text{avr}(s, t, u))$$

Algorithm 2: Authorization for Resources

A resource's entry request that a physical/virtual Resource r requests to enter a Secure Space s at a Time t is granted, if and only if any user inside there permits accessing the Resource and it permits being accessed by any user inside there in the context, that is, in the Secure Space which contains some set of entities such as visitors at the Time.

$$\forall r \in R, \forall s \in S, \forall t \in T, \ \text{authR}(r, s, t) = \text{grant}$$
$$\Leftrightarrow \text{au}(s, t, r) = \text{cu}(s, t)$$

where $\text{au}(s, t, r)$ is the Assumptive set of Users inside the Secure Space s after forcing all users who don't permit accessing the Resource r or it doesn't permit being accessed by to exit there in order to keep secure.

$$\text{au}(s, t, r) := \{u_k \in \text{cu}(s, t) | (u_k, r, s, t, \text{cu}(s, t)) \in \text{UAP}$$
$$\text{and } (r, u_k, s, t, \text{cu}(s, t)) \in \text{RAP}\}$$

5   Architecture

This section introduces a system architecture for our content-based entry control model to build Secure Spaces in the physical world. Each Secure Space requires the following facilities at least. Figure 8 gives an overview of our system.

- **Space Management:** is responsible for constantly figuring out its visitors and its contents such as physical resources and virtual resources via its embedded output devices, for ad-hoc making authorization decisions on whether an entry request to enter there by a user or a physical/virtual resource should be granted or denied, and for notifying the entry decisions to the Electrically Lockable Doors or enforcing entry control over virtual resources according to the entry decisions by itself.
- **User/Object Authentication:** is responsible for surely authenticating physical entities such as users and physical resources which requests to enter or exit the Secure Space by using Radio Frequency IDentification or biometrics technologies and for notifying it to the Space Management.
- **Electrically Lockable Door:** is responsible for electrically locking or unlocking itself in order to enforce entry control over physical entities such as users and physical resources according to the entry decisions notified by the Space Management.
- **Opaque and Soundproof Wall:** is responsible for physically isolating a Secure Space from outside with regard to information access in order to validate our assumption that any user inside a Secure Space can access any resource inside the Secure Space but any user outside the Secure Space can never access any resource inside the Secure Space.

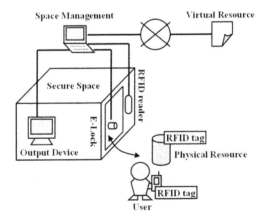

Fig. 8. An architecture for building Secure Spaces

6   Related Work

Our work in this paper and the previous ones [1,2] is related very well to the research field of authentication and access control for Smart Spaces.

Smart Spaces [3], also called Intelligent Environments [4], are physical spaces (often physically isolated, e.g., rooms or buildings) which heterogeneous computing resources such as output devices, multi-modal sensors or communication apparatus are embedded in and which provide various advanced services for their visitors by cooperating with their resources. They might have the ability to observe the physical world, to interpret the observation, to perform reasoning on the interpretation, and then to perform actions based on the reasoning. These examples include Active Space in Illinois [5], Interactive Workspace in Stanford [6], Aware Home in Georgia Tech [7], EasyLiving in Microsoft [8], Smart/Intelligent Room in MIT [9,10], Smart Classroom [11] and so forth.

As one of the hottest topics in such researches on Smart Spaces, it becomes emphasized that information environments which they provide for their visitors should be not only entertaining or convenient but also comfortable or secure. For example, Cerberus which is a context-aware security infrastructure using first-order predicate and boolean algebra for Smart Spaces [12], a role-based context-aware access control mechanism within Hyperglue for Intelligent Environments [13], a role-based access control system switching four modes of Individual, Shared, Collaborative, and Supervised by situation for Active Spaces [14], iSecurity which is a security framework for iRoom as an Interactive Workspace [15], and a security architecture using Environment Roles for Aware Home [16].

7   Conclusion and Future Work

This paper redefined Secure Spaces as physical environments which protect not only their contents' freedom of information delivery but also their visitors' freedom of information access, and proposed a unified model and an architecture for content-based entry control to make public/shared spaces more secure.

In the future, we plan to formalize a hierarchical and more flexible model and specify an access policy description language in order to allow space administrators to configure their Secure Space more easily and more flexibly, and what is more, we will implement a prototype of Secure Space and evaluate its effectivity or functionality by applying it to actual use cases in the real world.

Acknowledgments. This work was supported in part by the Japanese MEXT 21st Century COE (Center of Excellence) Program "Informatics Research Center for Development of Knowledge Society Infrastructure" (Leader: Katsumi Tanaka, 2002–2006), and the Japanese MEXT Grant-in-Aid for Scientific Research on Priority Areas: "Cyber Infrastructure for the Information-Explosion Era", Planning Research: "Contents Fusion and Seamless Search for Information Explosion" (Project Leader: Katsumi Tanaka, A01-00-02, Grant#: 18049041).

References

1. Hattori, S., Tezuka, T., Tanaka, K.: Content-Based Entry Control for Secure Spaces. In: Proc. of the International Workshop on Future Mobile and Ubiquitous Information Technologies (FMUIT'06) in conjunction with MDM'06, vol. 98 (2006)
2. Hattori, S., Tezuka, T., Tanaka, K.: Secure Spaces: Physically Protected Environments for Information Security. In: Proc. of the Joint 3rd International Conference on Soft Computing and Intelligent Systems and 7th International Symposium on advanced Intelligent Systems (SCIS&ISIS'06), pp. 687–691 (2006)
3. Rosenthal, L., Stanford, V.M.: NIST Smart Space: Pervasive Computing Initiative. In: Proc. of the 9th IEEE International Workshops on Enabling Technologies: Infrastructure for Collaborative Enterprises (WETICE'00), pp. 6–11 (2000)
4. Coen, M.H.: Design Principles for Intelligent Environments. In: Proc. of the 15th National/10th Conference on Artificial Intelligence/Innovative Applications of Artificial Intelligence (AAAI/IAAI'98), pp. 547–554 (1998)
5. Roman, M., Hess, C., Cerqueira, R., Ranganathan, A., Campbell, R.H., Nahrstedt, K.: Gaia: A Middleware Infrastructure to Enable Active Spaces. IEEE Pervasive Computing 1(4), 74–83 (2002)
6. Johanson, B., Fox, A., Winograd, T.: The Interactive Workspaces Project: Experiences With Ubiquitous Computing Rooms. IEEE Pervasive Computing 1(2), 67–74 (2002)
7. Kidd, C.D., Orr, R., Abowd, G.D., Atkeson, C.G., Essa, I.A., MacIntyre, B., Mynatt, E., Starner, T.E., Newstetter, W.: The Aware Home: A Living Laboratory for Ubiquitous Computing Research. In: Proc. of the 2nd International Workshop on Cooperative Buildings (CoBuild'99). LNCS, vol. 1670, pp. 191–198. Springer, Heidelberg (1999)
8. Brumitt, B., Meyers, B., Krumm, J., Kern, A., Shafer, S.: EasyLiving: Technologies for Intelligent Environments. In: Proc. of the 2nd International Symposium on Handheld and Ubiquitous Computing (HUC'00). LNCS, vol. 1927, pp. 12–29. Springer, Heidelberg (2000)
9. Pentland, A.: Smart Clothes, Smart Rooms. In: Proc. of WETICE'01, vol. 3 (2001)
10. Brooks, R.A.: The Intelligent Room Project. In: Proc. of the 2nd International Conference on Cognitive Technology (CT'97), pp. 271–278 (1997)
11. Xie, W., Shi, Y., Xu, G., Xie, D.: Smart Classroom - An Intelligent Environment for Tele-education. In: Proc. of the 2nd IEEE Pacific-Rim Conference on Multimedia (PCM'01). LNCS, vol. 2195, pp. 662–668. Springer, Heidelberg (2001)
12. Al-Muhtadi, J., Ranganathan, A., Campbell, R., Mickunas, M.D.: Cerberus: A Context-Aware Security Scheme for Smart Spaces. In: Proc. of the First IEEE International Conference on Pervasive Computing and Communications (PerCom'03), pp. 489–496 (2003)
13. Kottahachchi, B., Laddaga, R.: Access Controls for Intelligent Environments. In: Proc. of the 4th Annual International Conference on Intelligent Systems Design and Applications (ISDA'04) (2004)
14. Sampemane, G., Naldurg, P., Campbell, R.H.: Access Control for Active Spaces. In: Proc. of the 18th Annual Computer Security Applications Conference (ACSAC'02), pp. 343–352 (2002)
15. Song, Y.J., Tobagus, W., Leong, D.Y., Johanson, B., Fox, A.: iSecurity: A Security Framework for Interactive Workspaces. Tech Report, Stanford University (2004)
16. Covington, M.J., Long, W., Srinivasan, S., Dev, A.K., Ahamad, M., Abowd, G.D.: Securing Context-Aware Applications Using Environment Roles. In: Proc. of the 6th ACM Symposium on Access Control Models and Technologies (SACMAT'01), pp. 10–20 (2001)

Ventricular Tachycardia/Fibrillation Detection Algorithm for 24/7 Personal Wireless Heart Monitoring

Steven Fokkenrood*, Peter Leijdekkers**, and Valerie Gay**

University of Technology, Faculty of IT,
PO Box 123 Broadway NSW 2007, Australia
* s.a.w.fokkenrood@student.utwente.nl
** {peterl,valerie}@it.uts.edu.au

Abstract. This paper describes a Ventricular Tachycardia/Fibrillation (VT/VF) detection algorithm that is specifically designed for a 24/7 personal wireless heart monitoring system. This monitoring system uses Bluetooth enabled bio-sensors and smart phones to monitor continuously cardiac patients' vital signs. Our VT/VF algorithm is optimized for continuous real-time monitoring on smart phones with a high sensitivity and specificity. We studied and compared existing VT/VF algorithms and selected the one which suited best our requirements. However, we modified and improved the existing algorithm for the smart phone to achieve better performance results. We tested the algorithm on full-length signals from the physionet CU, MIT-db and MIT-vfdb databases [16] without any pre-selection of VT/VF or normal QRS-complex signals. We achieved 97% sensitivity, 98% accuracy and 98% specificity for our implementation which is excellent compared to existing algorithms.

Keywords: ventricular/tachycardia algorithm, ECG signal processing, heart monitoring, mobile-health, wireless ECG sensors.

1 Introduction

Cardiovascular diseases are now the number one cause of death in the United States and most European countries. The costs of medical help for cardiovascular patients is already estimated to be €316 billion in the United States in 2006 [1]. Using state-of-the-art sensor technology and smart phones, it is now possible to lower the costs related to heart patients. The latest ECG monitors are now wearable and reliable, allowing the patient to wear them 24/7. The use of mobile health applications to monitor patients is a booming business worldwide. The market is estimated to be $81 billion (USD) in the United States alone.

At the University of Technology, Sydney, we work on a context-aware personal health monitoring project [2]. In this project we use smart phones and wearable Bluetooth bio-sensors for patient monitoring and rehabilitation. When a life threatening situation is detected, the smart phone automatically calls and SMSes pre-programmed phone numbers indicating the location and reason for the emergency call. Also potential bystanders are notified by a message played continuously on the smart phone and instruct them what to do.

T. Okadome, T. Yamazaki, and M. Mokhtari (Eds.): ICOST 2007, LNCS 4541, pp. 110–120, 2007.
© Springer-Verlag Berlin Heidelberg 2007

This paper focuses on one fundamental part of this project: the detection of Ventricular Fibrillation (VF) which is the main cause of sudden cardiac death and Ventricular Tachycardia (VT). Ventricular Tachycardia is a very rapid beating of the heart, resulting in a diminished cardiac output. Our objective is to implement an efficient VT/VF detection algorithm for a mobile device which should satisfy the following requirements:

- It must be able to process the ECG data in real-time on the smart phone. To save battery power we need to avoid unnecessary processing on the smart phone.
- The algorithm should detect VT/VF with high sensitivity and specificity. We want to minimize the possibility of missing a VT/VF episode which can be lethal.
- Minimise false alarms. The classification of a QRS complex as VT/VF (i.e. false classification) is inconvenient for the patient using the monitoring system.
- We do not need to distinguish between VT and VF since it is not meant to be used in defibrillators to decide whether an arrhythmia is shockable.

Numerous algorithms exist to detect VT and VF and we evaluated them against our requirements. We also investigated the triggers for a VT/VF onset in order to improve the algorithm. Public online databases are used to test our algorithm and we compared the performance of our algorithm against other algorithms [3 - 9] using the same dataset.

This paper is organised as follows. Section 2 evaluates existing VT/VF algorithms and identifies those that are a good basis for our project. Section 3 discusses the design and implementation of our improved VT/VF algorithm. Section 4 presents the performance results and section 5 shows how the VT/VF algorithm is integrated in the smart phone application. Finally, section 6 concludes this paper.

2   Evaluation of Publicly Available VT/VF Algorithms

Two factors are of particular importance which are high *sensitivity* and *specificity*. The sensitivity is a measure for how accurate the algorithm detects VT/VF episodes in an ECG. Specificity corresponds to the algorithm's capability to alarm if, and only if, it is an actual VT/VF signal.

In addition *limited processing power* of the smart phone implies restrictions on the possible algorithms for implementation since the smart phone has limited processing capabilities (typically a 200-500 MHz processor). Algorithms that are not practical to implement on a mobile device are Fuzzy Neural Network (FNN) designs or Wavelet Transformation (WTF) based algorithms. A FNN algorithm is often implemented on high end processors which are not yet available for smart phones [4]. WTF algorithms are very CPU intensive which is not practical for mobile phones and will drain the battery quickly [4].

The smart phone provides an *ambulatory monitoring* environment for the patient. The patient is able to move freely while being continuously monitored. We selected a modified lead II type ECG sensor with only two electrodes attached directly to the sensor. Using a 2-lead sensor implies that we look at algorithms that use the physionet databases MLII signals and omitted those that use their own intracardiac ECG dataset [3].

Several algorithms claim high sensitivity and specificity, but on closer inspection many algorithms use pre-selections of ECG database signals to evaluate their performance [5] or pre-select MLII input signals [6-8]. Algorithms tested on partial signals are; Threshold Crossing Interval (TCI) [7], Complexity Measure (CM) [8], Modified Complexity Measure (MCM) [5] and the Auto-Correlation Function (ACF$_{95/99}$) [6]. ACF$_{95}$ and ACF$_{99}$ refer to Fisher statistics degree of freedom.

Table 1 shows the results of the algorithms on pre-selected signals as published in [5-8], as well as, the results of the same algorithms tested on full-length signals [4]. In case of full-length signals the performance of the algorithms drops significantly. The TCI algorithm is generally regarded as a good algorithm [4] for VT/VF detection and reaches 75.1% sensitivity. However, it classifies any heart rate higher than 150 bpm as VT or VF which is an incorrect assumption since a healthy person can easily have a heart rate of 150 bpm and higher when exercising.

Table 1. Performance results (7stw= 7-second time window)

Algo	Pre-selected signals		Full length signals (8stw, [4])	
	Sensitivity (%)	Specificity (%)	Sensitivity (%)	Specificity (%)
TCI	100 (7stw, [7])	N/A	75.1	84.4
CM	100 (7stw, [8])	100 (7stw, [8])	59.2	92.0
MCM	89.8 [5])	N/A	51.2	84.1
ACF$_{95}$	N/A	N/A	49.6	49.0
ACF$_{99}$	100 (4.5stw, [6])	100 (4.5stw, [6])	69.2	35.0
SPEC	93 (8stw, [9])	79 (8stw, [9])	29.1	99.9
VF	91	94	18.8	100
ADA			95.9 (10stw, [9])	94.4 (10stw, [9])
SCA			71.2 (8stw, [4])	98.5 (8stw, [4])
QRC			91.9 (6stw, [12])	98.3 (6stw, [12])

Jekova describes in [9] the performance results for the Spectral analysis (SPEC) [10] and VF-Filter (VF) [11] algorithms on pre-selected MLII signals. These algorithms perform very well with respect to specificity but sensitivity is very low for full-length signals for both algorithms.

The Amplitude Distribution Analysis (ADA) algorithm [9], Signal Comparison Algorithm (SCA) [4] and QRS detection and Rhythm Classification algorithm (QRC) [12] use full-length signals to test their performance. ADA applies noise detection, asystole detection, VT/VF detection and beat detection to classify a heart signal. SCA compares a heart signal interval with four different reference signals to classify a heart signal episode (i.e. three QRS-complexes and one VT/VF reference signal). The residual between the measured signal and the four reference signals is a measure for classification. ADA, SCA and QRC are tested on full-length signals from the MIT-db, MIT-vfdb, AHA and CU databases. The advantage of the ADA, QRS and SCA algorithms is the use of several steps to classify a heart signal. This improves the performance considerably compared to other algorithms.

From our literature study we found out that the SCA algorithm is not a practical solution for our application since it will be very hard to obtain a VT and VF reference signal for each patient. We agree with Fernandez about the QRC algorithm where he states that "*System failed, in general, when signal, even filtered, holds noisy and the noise peaks had very big amplitude"* [5, 12]. It will fail in an ambulatory environment for our 2-lead sensor. The ADA algorithm is the best candidate for implementation on the smart phone where beat detection and classification is performed together with VT/VF detection [5, 12].

3 Improved VT/VF Algorithm

We modified the original ADA algorithm and altered the order of the signal processing routines. We placed the beat detector before the VT/VF classification as suggested by Fernandez in [12]. This improves the overall performance of the algorithm since the beat detector is less CPU intensive compared to the VT/VF routine. Fig. 1 shows the improved ADA algorithm as implemented for the mobile device.

Beat detector
ECG data is collected from the 2-lead ECG sensor or physionet ECG files. We use the open source beat detector from Hamilton [13] to obtain the heart rate and beat type which can be QRS, Premature Ventricular Contraction (PVC) or Unknown. Based on the heart rate and/or beat type we check for VT/VF for the following conditions:

- Heart rate lower or higher than the threshold set by a cardiologist. The thresholds are patient specific and might indicate something abnormal for that patient [14].
- High Heart Rate Variability (HRV) can indicate the onset of VT and VF [14].
- High number of PVC's within a certain time limit can indicate a VT onset [14].
- Many 'unknown' beat types in a short time interval indicate that something is wrong with the heart signal because QRS-complexes cannot be detected.

6 second signal collection
The algorithm will collect 6 seconds of ECG data for further analysis if one of the above mentioned events occurs. From simulation studies conducted with Matlab we determined that 6 seconds is sufficient to analyse a signal efficiently. Signal classification is based on statistical distributions which needs a sufficient amount of signal samples. A longer time window forces the algorithm to classify a signal collection possibly containing both QRS and VT/VF signals as either QRS or VT/VF which is incorrect. On the other hand, a too short time window will result in the algorithm not being able to classify a signal correctly and will return an 'unknown' result. We also collect the heart rate, beat types and average heart rate for this 6 second interval.

QRS-complex and heartbeat classifier
This routine checks for high HRV and the number of PVCs or 'unknown' heart beats in the 6 second signal. Further processing of the signal is only required, if:

- The HRV remains high during this 6 second interval.
- The number of detected PVC's or 'unknown' beat types is higher than 50% of the detected beat types.

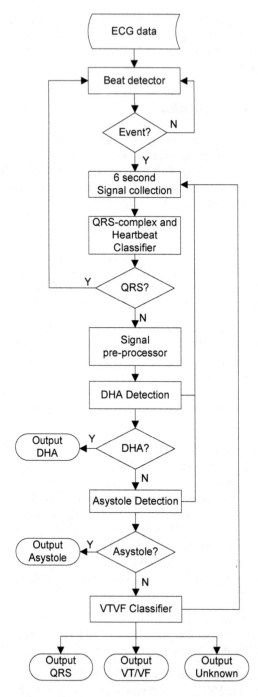

Fig. 1. Flow chart for the modified ADA algorithm

Signal pre-processor
In case of an expected arrhythmia the 6 second signal is filtered as follows:

- The signal is normalized, removing the gain of the signal. Gain removal allows millivolt (mV) thresholds for the DHA detector and Asystole classifier.
- The signal is then forward-backward filtered though a 1.4Hz high-pass filter. This removes the low frequency noise causing offset and drift of the ECG signal.
- The signal is then forward-backward filtered through a low-pass filter. This removes the high frequency noise from muscle movement using a 30Hz low-pass filter [3, 4, 12]. Forward-backward filtering of the signal ensures that there is no amplitude distortion and phase shifting of the signal.

DHA detection
We then check the signal for Dangerous Heart Activity (DHA). DHA can be a noisy ECG signal or an arrhythmic VT or VF signal. Large amplitude differences may imply a noisy signal but if there are too many oscillations it can be a VF signal since fast contractions of the ventricles result in high oscillating skew rates in the ECG signal. A higher amplitude of the current epoch compared to the previous 6 second epoch is used to ignore unreadable ECG signal periods. When DHA is detected the algorithm will automatically analyse the following 6 second epoch.

Asystole Detection
In the absence of large oscillations, the epoch is then analysed for a low amplitude signal. Asystolic signal cannot be analysed with the normal VT/VF classifier because of the low amplitude. If the amplitude is below 0.3mV we will classify the signal with the asystole classifier. If not, we use the VT/VF classifier to determine the type of heart signal.

Asystole Classifier
When an asystole is suspected we use an asystole classifier to determine the signal type (i.e. QRS, Asystole or VT/VF). We determine the following characteristics of the signal:

- We calculate the number of peaks in the signal. This is a rough indicator for the heart rate.
- The relation between the samples above and below the mean signal. For QRS signal this relation is lower than for VT/VF signal [12].
- The overall amplitude of the signal.

The signal is regarded asystolic when the maximum amplitude does not cross the 0.1mV threshold. The signal will be classified as a QRS complex if less than 40% of the signal is above half the maximum value, and the number of possible QRS complexes is below the emergency heart rate. Otherwise, it will be classified as a VT/VF signal. The Asystole classifier will trigger the algorithm to automatically analyse the following 6 second epoch.

VT/VF Classifier
When the signal epoch is not classified as asystole by the Asystole detector we then process it by the VT/VF classifier. This algorithm is based on statistical distribution

of QRS and VT/VF signal samples as proposed by Jekova [9]. For each 6 second interval (I), three amplitude ranges are defined, and signal samples (S$_i$) falling within a certain amplitude range for this interval are added, which are shown in equation 1.

$$T_1 = \left(0.8 \cdot max(|I|) \le S_i \le max(|I|)\right)$$
$$T_2 = \left(|\bar{I}| \le S_i \le 0.95 \cdot max(|I|)\right)$$

$$T_x = \left((|\bar{I}| - MD) \le S_i \le (|\bar{I}| - MD)\right), \text{ with } MD = \frac{1}{n}\sum|S_i - \bar{I}| \tag{1}$$

$$T_3 = \frac{T_1 \cdot T_2}{T_x}$$

The result of the VT/VF classifier depends on the signal samples accrued for T_1, T_2 and T_3, which are based on Jekova's algorithm [9]. Classification of the signal is presented in equation 2. An epoch processed by the VT/VF classifier will always trigger the algorithm to analyse the next 6 second epoch even if the result is a QRS complex.

$$QRS = (T_1 < 120 \ \& \ T_2 < 456 \ \& \ T_3 < 100) \ V$$

$$(120 < T_1 < 192 \ \& \ T_2 < 288 \ \& \ T_3 < 100)$$

$$VT/VF = (T_1 < 120 \ \& \ T_2 \ge 456 \ ) \ V \tag{2}$$

$$(T_1 \ge 120 \ \& \ T_3 \ge 100) V \ (T_2 \ge 528)$$

$$Unknown = Everything \ else$$

4 Performance Results

We implemented the algorithm on a Microsoft Windows Mobile Pocket PC platform using the C# programming language. The algorithm is tested on the following full-length signals from the online physionet databases [15, 16]:

- CU files: CU01-CU35.
- MIT-db files: 100-119, 121-124, 200-203, 205, 207-215, 217, 219-223, 228, 230-234.
- MIT-vfdb files: 418-430, 602, 605, 607, 609-612, 614, 615.

We did not modify or pre-select any of the signals. The MIT-db and MIT-vfdb files contain MLII and V5 signals but we only used the MLII signal since our ECG sensor is a 2-lead MLII type sensor. The results of the algorithm are represented by four parameters as shown in equation 3.

$$Se = \frac{TP}{TP + FN} \quad Sp = \frac{TN}{TN + FP} \quad PP = \frac{TP}{TP + FP} \tag{3}$$

$$Acc = \frac{TP + TN}{TP + FP + TN + FN}$$

- True positive (TP) is the correct classification of a VT or VF signal.
- True negative (TN) is the correct classification of a normal heart signal.
- False positive (FP) is the incorrect classification VT/VF while the heart signal is normal.
- False negative (FN) is the incorrect classification of a QRS complex while the heart is in a VT/VF state.

Sensitivity (Se) indicates the capability to detect all occurring VT/VF episodes whereas *Specificity (Sp)* signifies the discriminating quality of the algorithm.

Positive Predictivity (PP) indicates the probability that a signal epoch classified as VT/VF is truly VT/VF. It is a measure for the inconvenience to the patient, since it represents the ratio between actual VT/VF arrhythmia and false alarms. *Accuracy (Acc)* specifies all the correct decisions made by the algorithm.

The output of the algorithm can be QRS, DHA, Asystole, VT/VF or Unknown and the time of occurrence. We manually compared the output of the algorithm with the physionet ECG annotations. Automatic comparison is not possible since the format of the annotation files vary per database. In order to relate the annotations with the algorithm output we interpreted the algorithm output as follows:

- DHA output is classified as TP or FN. The reason being that DHA could be a dangerous rhythm.
- Unknown output is always an incorrect classification which results in FP for a QRS signal, and FN for a VT/VF signal.
- If the physionet annotations define a signal interval as noise, unreadable or asystole, the output of the algorithm cannot be classified as TP, FP, TN or FN. These heart signals do not contain useful information and therefore not contribute to the performance result of the algorithm.
- Asystole output does also mean that the heart signal does not contain useful information. The amplitude of the signal is too low to detect any QRS or arrhythmia characteristic. Therefore, this output is not classified as TP, FP, TN or FN.
- The algorithm can only analyse and classify 6 second intervals. If the physionet annotation corresponds with the output within this 6-second interval it is used as a valid result.

The results are presented in Table 2.

Table 2. Performance results

	CU	MIT-vfdb	MIT-db	Overall
Sensitivity	95%	91%	100%	97%
Specificity	96%	96%	99%	98%
Positive predictivity	81%	85%	65%	73%
Susceptibility	96%	96%	99%	98%
Accuracy	95%	96%	99%	98%

The overall sensitivity performance of 97% corresponds to the recognition of every single VT/VF epoch. But we actually detect 100% of the VT/VF episodes as long as they are longer than 6 seconds. If VT/VF episodes are shorter than 6 seconds we do not classify them as VT/VF since the heart returns to a normal state and in our situation we do not need to raise an alarm.

The overall specificity of 98% indicates that the algorithm will raise an alarm in case of a real VT/VF signal in 98% of the cases. The remaining few cases (2%) correspond to unreadable parts of a signal. If the signal is unreadable we should alarm the user so that he/she can check the ECG sensor.

Positive predictivity is not a good indication for the number of false alarms. A better indicator would be the amount of false positive classifications compared to the official number of QRS epochs. This indicator is called *Susceptibility* and corresponds to the amount of FP classifications during a QRS signal, see equation 4. It is a more accurate measure for the number of false alarms. If we apply this equation for our algorithm, we achieve 98% susceptibility.

$$Su = 1 - \frac{FP}{TN} \qquad (4)$$

5 Integration with the Personal Health Monitor Application

We integrated the algorithm with the personal health monitor application and tested its functionality using the MIT-db, MIT-vfdb and CU databases. The application has a demo-mode functionality where pre-recorded (physiobank) ECG files can be processed by the health monitor application. The application processes the physiobank signals and live ECG data in the same manner. The only difference is that in the demo-mode the actual alarm functionality (i.e. sending SMS and call) is disabled. In non demo-mode the application generates an alarm to notify the user

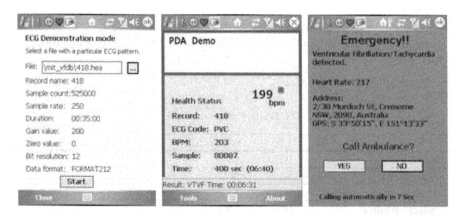

Fig. 2. Detecting VT/VF and raise an alarm

when the algorithm detects a dangerous arrhythmia (i.e. VT/VF, Asystole or DHA). If the user does not react in a certain time the application will automatically call and SMS pre programmed phone numbers stating the reason of the call and the current location of the patient.

We tested the performance of the algorithm on two different smart phones (I-mate® JASJAR, 520MHz processor and i-mate® K-JAM, 195MHz processor). Both smart phones are able to process the physionet ECG data files in real-time. Tests with the Alive ECG sensor showed no problem processing ECG data in real-time.

6 Conclusion

The heart monitoring algorithm implemented for the smart phone is able to detect VT/VF, Asystole and dangerous heart activity (DHA) with high sensitivity and specificity. We tested the algorithm on full-length signals without any pre-selection of VT/VF or normal QRS-complex signals. We achieved 98% accuracy, 97% sensitivity and 98% specificity which is an excellent score compared to existing algorithms and suitable for the personal health monitor application on the PDA.

We have also put mechanisms in place to deal with the few false alarms allowing the patient to disable the alarm. An ECG signal is recorded automatically as soon as it is identified as abnormal which allows further examination by the cardiologist. The patient can also start recording if he/she feels an abnormal heart rate or rhythm This means that our 24/7 Personal Wireless Heart Monitoring System can adequately be used to detect VT/VF episodes.

Time is a crucial factor when VT/VF is detected. For the patient to have a chance to survive VT/VF, a defibrillator should be applied within 5 minutes. Our algorithm detects the onset of VT/VF within 6 seconds. It will raise an alarm if another VT/VF epoch is detected to avoid false alarms. This means that after 12 seconds of a VT/VF onset, emergency services and caregivers are notified via SMS and automatically placed phone calls. This will increase the chance that help can be given in time.

Point of interest is when we want to raise an alarm. An alarm can be raised after the first VT/VF detection or additional VT/VF classifications can be required. This depends on the situation of the patient. For example, if the patient is in a nursing home we could alarm the medical staff as soon as we have a VT/VF epoch detection. If the patient lives alone, we would call the emergency services after two 6 second VT/VF epochs to reduce false alarms. Whatever the answer is, our application can be personalised to suit the situation.

The next step is to improve the algorithm to detect earlier signs of heart failure and other arrhythmia associated with cardiovascular diseases such as Atrial Fibrillation. This will allow us to identify signs of an upcoming cardiac arrest or identify abnormalities and therefore able to warn the patient or caregiver earlier.

References

1. AHA. Cardiovascular Disease Cost. Last accessed (April 3, 2007) [cited, Available from http://www.americanheart.org/]
2. Leijdekkers, P., Gay, V.: Personal Heart Monitoring and Rehabilitation System using Smart Phones (2006)

3. Throne, R.D., Janice, M., Jenkins, L.A., Dicarlo, A.: Comparison of Four New Time Domain Techniques for Discriminating Monomorphic Ventricular Tachycardia from Sinus Rhythm using Ventricular Waveform Morphology. IEEE Transactions on Biomedical Engineering 38, 561–570 (1991)
4. Amann, A.R.T., Unterkofler, K.: Reliability of Old and New Ventricular Fibrillation Detection Algorithms for Automated External Defibrillators. BioMedical Engineering Online 4, 1–23 (2005)
5. Ayesta, U.L.S., Romero, I.: Complexity Measure Revisited: A New Algorithm for Classifying Cardiac Arrhythmias. IEEE Explorer 2, 1589–1591 (2001)
6. Chen, S., Thakor, N.V., Mower, M.M.: Ventricular Fibrillation Detection by a Regression Test on the Autocorrelation Function. Medical and Biological Engineering and Computing 25, 241–249 (1987)
7. Thakor, N., Yi-Zheng Shu, V., Kong-Yan, P.: Ventricular Tachycardia and Fibrillation Detection by a Sequential Hypothesis Testing. IEEE Transactions on Biomedical Engineering 37, 837–843 (1990)
8. Zhang, X.-S., Yi-Sheng, Z., Nitish, V.: Thakor and Zhi-Zhong Wang, Detecting Ventricular Tachycardia and Fibrillation by Complexity Measure. IEEE Transactions on Biomedical Engineering 46, 548–555 (1999)
9. Jekova, I., Krasteva, V.: Real Time Detection of Ventricular Fibrillation and Tachycardia. Physiological Measurements 25, 1167–1178 (2004)
10. Barro, S., Ruiz, R., Cabello, D., Mira, J.: Algorithmic Sequential Decision-making in the Frequency Domain for Life Threatening Ventricular Arrhythmias and Imitative Artefacts: a Diagnostic System. Journal of Biomedical Engineering 11, 320–328 (1989)
11. Kuo, S.D.R.: Computer Detection of Ventricular Fibrillation. In: Computers in Cardiology, pp. 347–349. IEEE Computer Society, Washington, DC (1978)
12. Fernandez, A.R., Folgueras, J., Colorado, O.: Validation of a Set of Algorithms for Ventricular Fibrillation Detection: Experimental Results. In: Proceedings of the 2nd Annual International Conference of the IEEE EMBS, Mexico (2003)
13. Hamilton, P.S., Tompkins, W.J.: Evaluation of QRS Detection Algorithms Using the IBM PC. Engineering Medical Biological Society, Annual Conference of the IEEE, pp. 830–833 (1985)
14. Nemec, J., Hammill, S.C., Shen, W.K.: Increase in Heart Rate Precedes Episodes of Ventricular Tachycardia and Ventricular Fibrillation in Patients with Implantable Cardioverter Defibrillators: Analysis of Spontaneous Ventricular Tachycardia Database. Scientific Congress of NASPE 22, 1729–1738 (1999)
15. Goldberger, A.L., Amaral, L.A.N., Glass, L., Hausdorff, J.M, Ivanov, P.C., Mark, R.G., Mietus, J.E., Moody, G.B., Peng, C.-K., Stanley, H.E.: PhysioBank, PhysioToolkit, and PhysioNet Components of a New Research Resource for Complex Physiologic Signals. Circulation 101, e215–e220 (2000)
16. MIT-BIH. CU Ventricular Tachyarrhythmia Database (1992) [cited, Available from: http://www.physionet.org/]

homeML – An Open Standard for the Exchange of Data Within Smart Environments

C.D. Nugent[1], D.D. Finlay[1], R.J. Davies[1], H.Y. Wang[1], H. Zheng[1],
J. Hallberg[2], K. Synnes[2], and M.D. Mulvenna[1]

[1] School of Computing and Mathematics, University of Ulster, N. Ireland
{cd.nugent,d.finlay,rj.davies,hy.wang,h.zheng,
md.mulvenna}@ulster.ac.uk
[2] Luleå University of Technology, Sweden
{josef.hallberg,kare.synnes}@ltu.se

Abstract. This work describes a potential solution to the problems caused by the heterogeneous nature of the data which may be collected within smart home environments. Such information may be generated at an intra- or inter-institutional level following laboratory testing or based on in-situ evaluations. We offer a solution to this problem in the form of a system/application/format independent means of storing such data. This approach will inevitably support the exchange of data within the research community and form the basis of the establishment of an openly accessible data repository. Within this abstract we present the outline design of homeML, an XML based schema for representation of information within smart homes and through exemplars demonstrate the potential of such an approach. An example of the typical type of software browser required for the data representation is also presented.

Keywords: Smart environments, heterogeneous data, xml schema, data repository.

1 Introduction

As the population continues to grow and the percentage of the elderly within the population also increases, there is a growing demand to develop and deploy technical solutions within the home environment to support levels of independence. This has the inevitable and indeed desired effect of extending the period of time a person can remain in their own living environment prior to the potential need of institutionalisation. It can also offer a positive impact on measures associated with quality of life. Solutions and services which may be deployed can take the form of, for example, remote vital signs monitoring [1], activity tracking [2] and assistive technology to control ambient settings [3]. In addition, devices and sensors may have the ability to monitor and record a person's movement within a room, to asses pressure values to gauge if a person is sitting in a specific chair or even in bed, or have the ability to measure the flow of water within the bathroom and kitchen. These types of sensory information are complemented by information from a series of specific supporting devices and as such may be considered as the underlying technology, upon which services are deployed by healthcare professionals.

T. Okadome, T. Yamazaki, and M. Mokhtari (Eds.): ICOST 2007, LNCS 4541, pp. 121–129, 2007.
© Springer-Verlag Berlin Heidelberg 2007

Given the importance of this domain and the huge benefits to be gained it has attracted much attention in the past decade from both commercial, healthcare and research perspectives. The net result has been the largely independent development of a series of specialised services by differing organisations, resulting in the generation of a multitude of data, which in effect is largely heterogeneous. Analysis of this data, depicting how users interact with the technology and the environment, provides useful information relating to lifestyle trends and how the environment may be adapted to improve the user's experience [4].

Taking the issues associated with the heterogeneous nature of the storage and the distribution of the data into consideration there is a growing need and indeed opportunity within the general smart environment research community to promote the development of a cross-system standard to support information exchange and improve accessibility to and analysis of the collected data. Adoption of such an approach could be deployed at both intra- and inter-institutional levels. Historically, the storage and distribution of information related to a person's activities within smart environments has been based on different formats and generally developed on a project-by-project basis. The research presented in this paper seeks to develop an XML-based system/application/format independent model, referred to as homeML, for representing and exchanging diverse data recorded within a range of service scenarios, monitoring devices and in-situ evaluations within the home environment.

2 Background

Although research efforts within the domain of smart homes have been prolific in the past decade it is becoming increasingly evident that there does not exist any centrally and freely available data repository of recorded information or open standards/protocols to follow in the acquisition and storage of data in such environments. This may hinder the exchange of data at both intra- and inter-organisational levels and its ensuing analysis. Efforts have been made to address this issue for complex physiologic signals through the commonly used PhysioNet resource [5]. Another example is the OpenECG portal [6] which aims to promote the consistent use of format and communication standards for electrocardiograms. Similar resources of this nature would be greatly beneficial within the domain of smart environments.

The issue and need to develop an open standard for the exchange of data should not be confused with the already existing standards and protocols which facilitate the seamless integration of numerous devices and services within the one environment. These efforts have strived to address the problems associated with compatibility issues between differing types of devices. Two of the most prominent standards which have emerged within this domain are the Open standards gateway initiative (OSGi) [7] and Konnex [8]. The motivation behind these standards has been to make all aspects of the deployment of home based technology simpler. This includes the connection of systems, the transfer of data and the management and provisioning of network devices. It is upon these services that the requirements to specify the structure of the data to be stored, as proposed by homeML, rests.

Efforts have also been focused towards the development of standards to support the data exchange of patient information in the form of electronic patient records (EPRs) on both intra- and inter-institutional levels. The most notably work in this domain can be found associated with the efforts of HL7 [9]. The mission of HL7 is to create standards for the exchange, management and integration of electronic healthcare information. Since 1997 HL7 has been recognised as the standard for electronic exchange of historical and administrative data in health services world wide. Although having widespread acceptability such standards do not facilitate the establishment of a data structure for data within smart environments, its recording, exchange or storage.

One of the key technology recommendations within HL7 is the use of the eXtensible Markup Language (XML) [10]. XML has gained much attention in recent years and of particular relevance has been described as an acceptable means to represent biomedical data. XML has also become a widely established solution for the expressing of a syntax for data and the exchange of data via the Internet. The main benefits of employing XML relate to its platform, vendor and application independence. It is also considered to offer an easy to follow hierarchical data structure. Within the general area of home based healthcare monitoring a limited number of studies have reported application of this technology. Wang et al. [11] have demonstrated how problems with the exchange of electrocardiogram data can be addressed via an open standard development based on XML. Feki et al. [12] have presented models for the development of XML based objects to support interaction between modules in environments offering support for those with disabilities. Finally, Ohmori et al. [13] have presented an XML based multimedia data acquisition and inquiry system for wearable computers. Although these studies have been successful in their own application domains, the issues of addressing the needs of a truly open standard for the exchange of data within smart environments has not been addressed. Through the use of such a standard it will become possible that all those involved within the research domain will have the opportunity to record and store their data in a common format which will support the establishment of a web based repository which can be freely accessible. In addition, flexibility in such a standard will allow it to evolve as the needs of users and their further understanding of smart environments develops. It is the applicability of XML to address the problems of data representation and storage within smart environments that we explore further in the following Sections.

3   Methods

As previously stated, there is a growing need to address the heterogeneity of the data generated within smart environments in addition to including necessary information such as patient details, any clinical interventions and details of any vital signs which may also have been recorded. In the first instance we have designed homeML as a first attempt to offer a solution for an open standard for the storage and exchange of data within smart environments. Our first step in this process has been to design homeML as a series of hierarchical data trees (Figure 1). We have supplemented this graphical representation with Tables 1 and 2 which describe the elements and attributes within the aforementioned models.

Each smart home starts with a root element *SmartHome*, which is uniquely identified by its attribute *homeID*. The main components for each smart home are *DateRecordCreated, SmartHomeInfo, Room, InhabitantDetails* and *Comment*. Each smart home can only have one set of each component aside from *Room* elements, of which a multiple number are allowed. It should be noted that, in line with practical deployment of smart homes, we have limited this schema to single occupancy only. It is expected that issues pertaining to multiple occupancy will be addressed in following versions. In the following paragraphs we provide a brief description of each of the elements included in the tree structure.

DateRecordCreated contains the date the XML file in question was created. *SmartHomeInfo* contains information on the location, region and details of those responsible for the smart home. This is contained within the elements *Address, Jurisdiction* (regional funding authority), *CareProvider* (care agency responsible for the home) and *ContactPerson* (individual immediately responsible). This is information common to the smart home which does not differ between rooms and is important especially in the instances of exchanging data between organisations.

Room as the name suggests represents a room in the smart home. There can be a number of rooms within the smart home and these are defined using the elements as shown in Figure 1. These elements include details of when the data was recorded (*DateOfRecording*), a *RoomDescription* (e.g. bedroom, bathroom, kitchen, etc), the *Floor*, which specifies the floor (level) which the room is on. There are two further elements which deserve further attention. The first of these is *RoomParam* which describes the basic make up of the room. This is represented using five further elements which specify the number of doors and windows and the size of the room (based on X, Y and Z values), initially we assume a room is square. The final element within *Room* is the *Device*, of which there can be any number. This element will hold information relating to devices that are fixed in each room. Typical examples of such devices would be motion detectors, water flow sensors, temperature sensors and door sensors. The *Device* is made up of several elements which are described in Table 2.

InhabitantDetails provides personal information relating to the person who is living within the home environment and their Care Plan. Depending on the ethical regulations at an institutional level, varying amounts of this information may be considered as optional. In addition, from a data exchange perspective, this information will not be made available and will be managed according to each institution's ethical guidelines. *InhabitantDetails* also contains *AmbulatoryDevices* which provides details on wearable health systems (e.g. Holter monitors and vital signs equipment) that the person may require to monitor their health status. The make up of *AmbulatoryDevices* (its sub-elements) is similar to that of *Device* which appears as an element of *Room*. The subtle differences are the optional elements *QuantizationResolution* (e.g. 1 volt), and *ElectrodePlacement* (general area or body part where electrode is placed) which replaces the location. Finally, *Comment* as the name suggests can contain non-specified information about the file.

Table 1. Description of SmartHome element based on the hierarchical representation of homeML in Figure 1

SmartHome	The root element for XML based smart home data			
Element/ Attribute	Description	Required/ Optional	Data type	Example
homeID	A unique identifier for the SmartHome	Required	Integer	23165146
DateRecordCreated	The creation date/ time of the record	Required	Date	2007/01/ 10T12:30:00
SmartHomeInfo	General information about the Smart home	Required	*Requires element description*	
InhabitantDetails	Personal information on inhabitants	Required	*Requires element description*	
Room	Definition of a room within the smart home	Required	*Requires element description*	
Comment	Additional comments about the smart home	Optional	String	Monitors glucose level and temp.

Table 2. Description of Device element based on the hierarchical representation of homeML in Figure 1

Device	The Device element for XML based smart home data			
Element/Attribute	Description	Required/ Optional	Data type	Example
deviceID	A unique device identifier	Required	Integer	13165146
Description	General description of the device	Optional	String	The device measures the room temp.
DeviceLocation	Details the	Optional		
XPos	absolute		float	4.5
YPos	coordinates of the		float	1.9
ZPos	device		float	6.2
DeviceType	Details the type of device	Required	String	Temp Sensor
Units	Measurement unit	Required	String	Celsius
RealTimeInfo	Specifies sample	Optional		
runID	rate, as well as		int	25341
SampleRate	start and end time,		float	10.1
StartTime	identified by the		Date	
EndTime	runID attribute		Date	
Event	Describes an	Optional		
eventID	event from the		int	64721
TimeStamp	device		Date	
Data			String	175

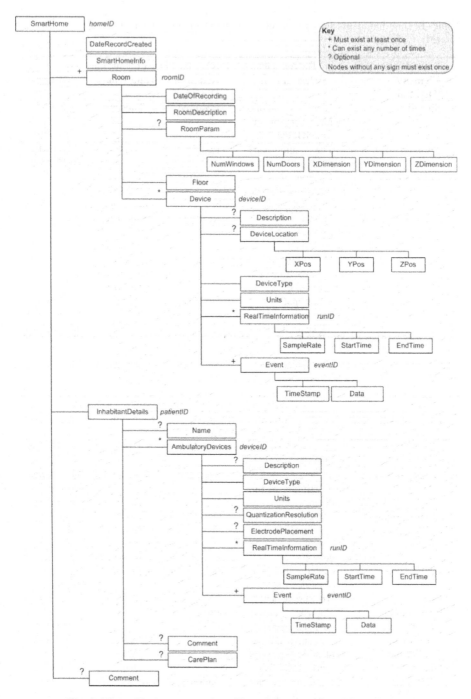

Fig. 1. Hierarchical representation of homeML as a series of Tree diagrams

4 Evaluation of the Model

Based on the design of homeML we have now produced the first version of the XML based schema. With this it is now possible to deploy a common standard at both intra- and inter-institutional levels for the recording of information collected within smart environments (either laboratory based experiments or real living environments)

```
<div class="moz-text-flowed" style="font-family: -moz-fixed"><?xml version="1.0" encoding="UTF-8"?>
<SmartHome xmlns:xsi="http://www.w3.org/2001/XMLSchema-instance"
xsi:noNamespaceSchemaLocation="homeML-v1.0.xsd" homeID="10000010">
  <DateRecordCreated>2007-01-04T12:30:00.05</DateRecordCreated>
  <SmartHomeInfo>
    <Address>16J27 </Address>
    <Jurisdiction>Community and Hospitals Trust</Jurisdiction>
    <CareProvider>Medical science</CareProvider>
    <ContactPerson>Dr. Martin</ContactPerson>
  </SmartHomeInfo>
  <Room roomID="301">
    <DateOfRecording>2007-01-04T16:35:45.27</DateOfRecording>
    <RoomDescription>bed room</RoomDescription>
    <RoomParam>
      <NumWindows>2</NumWindows>
      <NumDoors>1</NumDoors>
      <XDimension>12</XDimension>
      <YDimension>10</YDimension>
      <ZDimension>2.5</ZDimension>
    </RoomParam>
    ..............................
    ..............................
    ..............................
</SmartHome>
</div>
```

Figure (a)

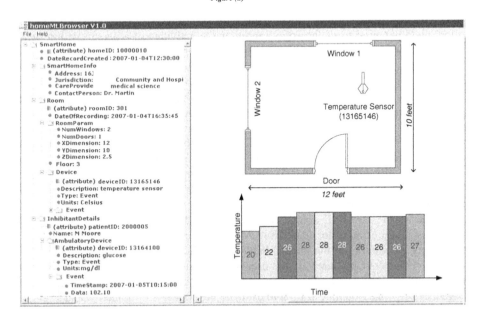

Figure (b)

Fig. 2. Overview of homeML concepts (a) excerpt of exemplar data based on the homeML XML schema (b) visions of prototype homeML browser to support data representation

which will facilitate the open exchange of data and will assist in moving one step closer towards the development of a common data repository for use by the entire research community. To facilitate uptake of this approach we will make available, through a series of web based resources, the XML schema along with documentation providing guidance on use and steps to follow to make recommendations for changes to the current version (http://trail.ulster.ac.uk/HomeML/).

In addition to providing the schema for homeML we are also planning to make available a suite of open source tools to assist users in exploiting the general concepts. In the first instance this will take the form of a viewer or browser which will provide a means to display the data recorded. Figure 2 provides an example of our vision for the development of such a tool. Figure 2 (a) provides an exemplar of an excerpt of data according to the homeML schema. Figure 2 (b) provides a prototype representation of how we envisage the homeML browser to be represented. On the left hand side we can view the tree structure (which can be mapped to our design in Figure 1). This can be expanded and collapsed at any level. The interface on the right hand side will have the ability to represent the layout of the room, the devices which have been included in the room and also to plot any time series data which may have been recorded. In this example, the browser is providing a means to plot temperature values within the room.

5 Conclusion

The ability to embed sensors within the home environment coupled with the increasing prevalence of wearable health care solutions for vital signs monitoring offers huge potential to unobtrusively monitor the activities of a person within their home environment and subsequently adapt the environment to offer a personalised living space. The types of devices/sensors/services which may be present within the home environment and the data they generate vary widely. It has been the aim of the current study to address the problems caused by the heterogeneous nature of the data generated within smart homes through the development of an open standard for data storage and exchange. These developments have been realised through the presentation of homeML, an XML based schema. This approach provides a common platform for the exchange of data in addition to offering suggestions for an environment for representation of the data through the proposed open source homeML browser. Developments are currently underway to develop the first version of this open source browser.

Access to the homeML source files will be made available through web based resources (http://trail.ulster.ac.uk/HomeML/) or can be obtained directly by contacting the authors of the paper. With this we hope to inform the community of these developments and maintain the user-driven notion of using XML. Finally, it is our intention, in the first instance, to secure collaborations with a small number of organisations who would be willing to evaluate the concepts of homeML and begin to establish a repository which would become widely available within the general research community. This will allow us to ascertain if homeML in its current form contains all the necessary information required to support meaningful analysis of persons within smart home environments.

References

1. Dittmar, A., Axisa, F., Delhomme, G., Gehin, C.: New concepts and technologies in home care and ambulatory monitoring. In: Lymberis, A., de Rossi, D. (eds.) Wearable eHealth Systems for Personalised Health Management: State of the Art and Future Challenges, pp. 9–35. IOS Press, Amsterdam (2004)
2. Philipose, M., Fishkin, K.P., Perkowitz, M., Patterson, D.J., Fox, D., Kautz, H., Hahnel, D.: Inferring activities from interactions with objects. IEEE Pervasive Computing, pp. 50–57 (2004)
3. Helal, S., Mann, W., El-Zabadani, H., King, J., Kaddoura, Y., Jansen, E.: The gator tech smart house: a programmable pervasive space. IEEE Computer, pp. 64–74 (2005)
4. Pollack, M.: Intelligent technology for an aging population: the use of ai to assist elders with cognitive impairment. AI Magazine, pp. 1–27 (2005)
5. PhysioNet, The research resource for complex physiologic signals http://ww.physionet.org
6. Open ECG Portal http://www.openecg.net/
7. OSGi - The dynamic module system for Java http://ww.osgi.org
8. Konnex - Open standard for home and building control http://ww.konnex.org
9. Health Level Seven http://ww.hl7.org
10. Extensible Markup Language (XML) http://www.w3.org/XML/
11. Wang, H., Azuaje, F., Jung, B., Black, N.: A markup language for electrocardiogram data acquisition and analysis (ecgML). BMC Medical Informatics and Decision Making 3 (2003)
12. Feki, M., Abdulrazak, B., Mokhtari, M.: XML modelisation of smart home environment. In: Proceedings of the 1st International Conference on Smart homes and health Telematics, pp. 55–60 (2001)
13. Ohmori, Y., Ouchi, K., Hattori, M., Doi, M.: An xml based multimedia data acquisition and data retrieval with wearable computers. In: Proceedings of the 21st International Conference on Distributed Computing Systems Workshop, pp. 272–277 (2001)

Characterizing Safety of Integrated Services in Home Network System

Ben Yan[1], Masahide Nakamura[1], Lydie du Bousquet[2],
and Ken-ichi Matsumoto[1]

[1] Nara Institute of Science and Technology (NAIST)
8916-5, Takayama-cho, Ikoma-shi, Nara, 630-0192 Japan
{hon-e,masa-n,matumoto}@is.naist.jp
[2] LSR Laboratory, IMAG, Joseph Fourier University (Grenoble I)
BP72, F-38402, Saint-Martin d'Hères Cedex, France
Lydie.du-Bousquet@imag.fr

Abstract. This paper formalizes three kinds of safety to be satisfied by networked appliances and services in the emerging home network system (HNS). The *local safety* is defined by safety instructions of individual networked appliances. The *global safety* is specified as required properties of HNS services, which use multiple appliances simultaneously. The *environment safety* is derived from residential rules in home and surrounding environments. Based on the safety defined, we propose a modeling/validation framework for the safety. Specifically, we first introduce an object-oriented modeling technique to clarify the relationships among the appliances, the services and the home (environment) objects. We then employ the technique of *Design by Contract with JML (Java Modeling Language)*, which achieves systematic safety validation through testing.

1 Introduction

The recent ubiquitous/pervasive technologies allow general household appliances to be connected within the network at home. The *home network system* (HNS, for short) is comprised of such networked appliances to provide various services and applications for home users[7]. The great advantage of HNS lies in *integrating* (or orchestrating) features of multiple appliances, which yields more value-added and powerful services. We call such services *HNS integrated services*. For example, integrating a TV, a DVD player, lights, sound-systems and curtains implements a *DVD Theater service*, which allows a user to watch movies in a theater-like atmosphere just within a single operation.

In developing and providing a HNS integrated service, the service provider must guarantee that the service is *safe* for inhabitants, house properties and their surrounding environment. In the conventional situations where a user operates (non-networked) appliances one-by-one, the safety has been assured manually by the human user. That is, every user is supposed to follow *safety instructions* typically described in a user manual.

T. Okadome, T. Yamazaki, and M. Mokhtari (Eds.): ICOST 2007, LNCS 4541, pp. 130–140, 2007.

On the other hand, as for the HNS integrated services, we have to consider the safety more carefully. Since the service is typically implemented as a software application, appliances are often operated automatically by the application, but not by the human user. Also, one integrated service operates multiple appliances, which yields global dependencies among different appliances. Moreover, since multiple integrated services can be executed, unexpected functional conflicts may occur among the services. Thus, a single fault in the service application can cause serious accidents to the user. Unfortunately, no solid study has been reported for the safety of HNS integrated services.

In order to achieve the safety within the HNS integrated services systematically, this paper formalizes three kinds of safety: (a) *local safety*, (b) *global safety*, and (c) *environment safety*. The *local safety* is defined by the safety instructions of individual networked appliances. The *global safety* is specified as required properties of HNS services, which use multiple appliances simultaneously. The *environment safety* is prescribed as residential constraints and rules in home and surrounding environments.

Based on the safety formulated, we then propose a modeling/validation framework. Specifically, we introduce an object-oriented modeling technique to clarify the relationships among the appliances, the services and the home (environment) objects. We then employ the technique of *Design by Contract* [6] with *JML (Java Modeling Language)* [3,9]. The properties of local, global and environment safety are represented as JML contracts, and embedded in Java source code of the appliance, the service and the home objects, respectively. Finally, the safety properties are validated through testing using related testing tools. Assuring safety is a crucial issue to guarantee high quality of life in smart home. We believe that the proposed framework provides a systematic approach to the safety assurance in the context of pervasive computing in smart home.

2   Preliminaries

2.1   Home Network System

A *home network system* (HNS) consists of one or more networked appliances connected within a LAN at home. In general, each appliance has a set of *application program interfaces* (i.e., APIs), by which the users or external software agents can control the appliance via the network. A HNS typically has a *home server*, which manages all the appliances in the HNS. Services and applications are installed on the home server. A *HNS integrated service* operates different multiple appliances together, and achieves a sophisticated and value-added service. An integrated service is implemented as a software application that invokes APIs of the appliances. It is supposed to be installed in the home server.

2.2   Example of HNS Integrated Services

We here introduce four example scenarios of HNS integrated services.

```
Public   DVDTheaterService {
         DigitalTV    tv = new DigitalTV();
         DVDPlayer  dvd = new DVDPlayer();
         SoundSystem  sound =new SoundSystem();
         Light   light = new Light();
         Curtain  curtain = new   Curtain();

         tv.on();                          /* Turn on TV */
         tv.setVisualInput('DVD');

         dvd.on();                         /* Turn on the DVD Player */
         dvd.setSoundOutput('5.1');

         sound.on();                       /* Turn on the Sound System */
         sound.setInputSource('DVD');
         sound.setVolumeLevel(25);

         curtain.closeCurtain();           /*  Close curtain  */

         light.setBrightnessLevel(1);      /*  Minimize brightness  */

         tv.playTv();                      /*  Play TV   */
         dvd.playDvd();                    /*  Play DVD  */
    }
```

Fig. 1. DVD theater service

[SS1: DVD Theater Service] Integrating a TV, a DVD player, a sound system, a light and a curtain, this service automatically sets up the living room in a theater configuration. Upon a user's request, the TV is turned on with the DVD input, the curtains are closed, the sound system is configured for 5.1ch mode, the light becomes dark, and finally the DVD player plays back the contents.

[SS2: Relax Service] Integrating a DVD player, a sound system, a light, an air-conditioner, and an electric kettle, this service helps a user relax in the living room. When the user starts the service, the DVD player is turned on with a music mode, a 5.1ch speaker is selected with an appropriate sound level, the brightness of the light is adjusted, the air-conditioner is configured with a comfortable temperature, and the kettle is turned on with a boiling mode to prepare hot water for coffee.

[SS3: Shower Service] Integrating a gas-boiler, a shower valve and an air-conditioner in the bathroom, this service provides a comfortable setting for taking shower. Upon the service request, the gas-boiler is turned on with a preset water temperature. When the user enters the bathroom, the shower valve is automatically opened to turn on the shower. Also, the air-conditioner is turned on to keep the user warm.

[SS4: Cooking Preparation Service] Integrating a gas-valve, a ventilator, a roaster and a kitchen light, and an oven, this service automatically sets up the kitchen configuration of preparing for cooking. When requested, the kitchen light is turned on, the gas-valve is opened, the ventilator is turned on, and the pre-heating of the oven is started.

Fig. 1 shows a Java-like pseudo code which implements the scenario SS1 of DVD Theater service. In the figure, X.Y() means the invocation of API Y() of appliance X.

3   Formalizing Safety of HNS Integrated Services

3.1   Safety of HNS

For a HNS integrated service, we define the safety in the *broad sense* as follows.

Definition 1 (safety in broad sense). A HNS integrated service s is *safe* iff s is free from any condition that can cause [injury or death to home users and neighbors], or [damage to or loss of home equipments and the surrounding environment].

Our long-term goal is to establish a solid framework that guarantees the safety in Definition 1. In general however, it is quite difficult to achieve the 100% safety. Hence, the safety is often evaluated by means of *risk*. Thus, to assure the safety to a considerable extent, a set of conditions or guidelines minimizing the risk (called, *safety properties*) are usually considered [2]. In the following subsections, we investigate the safety properties specific to the domain of the HNS and the integrated services.

3.2   Local Safety Properties for Individual Appliances

For every electric appliance, the manufacturer of the appliance prescribes a set of *safety instructions* for proper and safe use of the appliance. Conventionally, these instructions have been designated for human users. However, in the HNS integrated service, the instructions must be guaranteed within the software. For instance, the following shows a safety instruction for an electric kettle.

L1: Do not open the lid while the water is boiling, or there is a risk of scald.

Any integrated service using the kettle (e.g., SS2 in Section 2.2) must be implemented so that the service never opens the lid while the kettle is in the boiling mode. Other safety instructions include the installation issues. That is, every appliance must be installed in a proper environment described by its *specification*, including power voltage, rated current, power consumption, allowable temperature and humidity, etc.

Note that the safety instructions are a set of properties that are *locally* specified for each appliance. Thus we regard them as *local safety properties*. We assume that the local safety properties for an appliance are *determined by the vendor of the appliance*.

3.3   Global Safety Properties for Integrated Services

Since an integrated service orchestrates different multiple appliances simultaneously, it is necessary to consider *global properties* over the multiple appliances. For instance, SS3:Shower Service in Section 2.2 should guarantee the following safety property to prevent the user from getting scald or heart attack.

G1: When the service turns on the shower valve, the water temperature of the gas-boiler must be between 35 and 45 degree.

The next example shows a safety property for SS4:Cooking Preparation Service, which avoids carbon monoxide poisoning.

G2: While the gas valve is opened, the ventilator must be turned on.

Note that each of the properties is *globally* specified over multiple appliances. These *global safety properties* are usually service-specific, and are not covered by the local safety properties of individual appliances. Therefore, we suppose that the global safety properties are carefully *specified by the provider of the integrated service*.

3.4   Environment Safety Properties for House

In general, each house has a set of residential rules for inhabitants and neighbors to make a safe living. Since the integrated services give various impacts against the surrounding *environment* (including the room, the building, the neighbors, etc), the services must be safe against the environment by conforming to the residential rules. For instance, most house has a capacity of electricity, which yields the following safety property.

E1: The total amount of current used simultaneously must not exceed 30A.

Also for emergency, the following safety property should be concerned.

E2: Do not lock doors and windows in case of fire.

The following property may be derived from community rules.

E3: Do not make loud voice or sound after 9 p.m.

We assume that these safety properties are derived from the residential rules, including the house manual, the emergency procedure, community rules and policies, etc. We call such properties *environment safety properties*. Note that the environment safety properties are specified *independently* of appliances or services deployed in the HNS.

3.5   Safety Definition of HNS Integrated Services

Based on the discussion above, we define three kinds of safety as follows.

Definition 2 (safety of integrated service). Let s be a given integrated service, and

- let $App(s) = \{d_1, d_2, ..., d_n\}$ be a set of networked appliances used by s,
- let $LocalProp(d_i) = \{lp_{i1}, lp_{i2}, ..., lp_{im}\}$ be a set of local safety properties derived by the safety instructions of d_i,

- let $LocalProp(s) = \cup_{d_i \in App(s)} LocalProp(d_i)$,
- let $GlobalProp(s) = \{gp_1, gp_2, ..., gp_k\}$ be a set of global safety properties prescribed by s,
- let $EnvProp(s) = \{ep_1, ep_2, ..., ep_l\}$ be a set of environment safety properties derived from the environment where s is provided. Then,

Local Safety: s is *locally safe* iff s satisfies all properties in $LocalProp(s)$.
Global Safety: s is *globally safe* iff s satisfies all properties in $GlobalProp(s)$.
Environment Safety: s is *environmentally safe* iff s satisfies all properties in $EnvProp(s)$.
Safety: We say that s is *safe* iff s is locally, globally and environmentally safe.

We are now ready to formulate the safety validation problem.

Definition 3 (safety validation problem)

Input: A HNS h, an integrated service s, $LocalProp(s)$, $GlobalProp(s)$, and $EnvProp(s)$.
Output: A verdict whether s is safe or not within h.

4   Object-Oriented Modeling for Safety Validation

To conduct the safety validation problem systematically, this section presents an object-oriented model for the HNS and integrated services. Every networked appliance has the internal state (power status, driving mode, etc.) and the operational interfaces (i.e., APIs). Hence, it is reasonable to model each appliance as an *object* consisting of *attributes* and *methods*. Previously [5,8], has developed an object-oriented model for networked appliances. In addition to the appliance object, here we newly introduce a *service object* for the integrated service and a *home object* for the home environment.

Fig. 2 shows the overview of the proposed model described as a UML class diagram. The model mainly consists of three kinds of objects (classes): `Appliance`, `Service`, and `Home`. As specified in the diagram, these classes forms the following relationships: [R1: a `Home` has multiple `Appliances`], [R2: a `Home` has multiple `Services`], and [R3: a `Service` uses multiple `Appliances`]. These relationships match well the intuition of the HNS and integrated services.

4.1   Appliance Object

An appliance object models a networked appliance. The proposed model involves a super class `Appliance` and concrete appliance classes that inherit `Appliance`. The `Appliance` aggregates attributes and methods commonly contained in any kinds of electric appliances. It also has a `Specification`, which stores static specification information such as power voltage, rated current, size, allowable temperature and humidity. Typical methods involve the power switch (`on()`, `off()`), acquisition of current power consumption (`getCurrentConsumption()`), and query for the specification (`getApplianceSpecification()`).

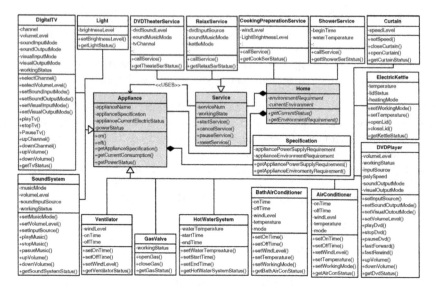

Fig. 2. Object-oriented model of HNS

On the other hand, operations specific to each kind of appliance are specified in the individual sub-classes. Such methods include `TV.selectChannel()`, `DVD.playDvd()` and `Kettle.openLid()`. When a method of an appliance is executed, values of some attributes are changed, which updates the *current state* (i.e., the tuple of the current values of all attributes) of the appliance. Preferably, every appliance should have a mothod such as `TV.getTvStatus()` so that the current state can be referred by external objects.

4.2   Service Object

A service object models an integrated service, which uses several appliance objects depending on contents of the service. Similar to `Appliance`, there is a super class `Service` which aggregates common operations such as `startService()` and `cancelService()`. The concrete service scenarios are specified in sub-classes that inherit `Service`. For instance, `DVDTheaterService` (see SS1 in Section 2.2) uses appliances `DigitalTV`, `DVDPlayer`, `SoundSystem`, `Light`, and `Curtain`.

4.3   Home Object

A home object models the house that involves environmental attributes, which is represented as a singleton object `Home`. The attributes of `Home` include the current energy consumption, sound level, brightness, temperature and humidity. We assume that these attributes can be obtained or computed from the current states of appliances and services. For instance, the current temperature is suppose to be obtained via `Home.currentEnvironment.getTemperature()`. The current electricity consumption is supposed to be computed from specifications and states of appliances that are currently on.

5   Safety Validation Framework with Design by Contract

5.1   Key Idea: Using Design by Contract (DbC)

In order to achieve the safety validation, we apply a software design strategy, called *design by contract* (DbC, for short) [3,6], to the object-oriented model presented in Section 4. For a given program, the DbC describes properties, conditions and invariants as a set of *contracts between calling and called objects*. The contracts are verified during runtime of the program under testing. During the execution, if a contract is violated, an exception is thrown or an error is reported. There are three kinds of contracts in the DbC.

Pre-Condition: A pre-condition of a method m is a condition that must be satisfied *before* executing m, which characterizes a premise of m.

Post-Condition: A post-condition of a method m is a condition that must be satisfied *after* executing m, which characterizes a consequence of m.

Class Invariant: A class invariant of a class c is a condition that must be guaranteed (i.e., kept unchanged) no matter which methods in c are executed.

Our key idea is to cope with the safety validation problem (see Definition 3) by first describing $LocalProp(s)$, $GlobalProp(s)$ and $EnvProp(s)$ as the DbC contracts, and then embedding them into the proposed object-oriented model. For this, we must consider carefully which object (`Appliance`, `Service`, or `Home`) should be responsible for $LocalProp(s)$, $GlobalProp(s)$ and $EnvProp(s)$.

5.2   Describing Local Safety Properties

Since the local safety properties are defined for individual appliance, `Appliance` or its sub classes should be responsible for $LocalProp(s)$. For instance, the local safety property L1 in Section 3.2 should be specified in `ElectricKettle` class, which can be encoded as the following contract:

```
Contractor:     ElectricKettle.openLid() method
Pre-condition:  heatingMode != 'boiling'
Post-condition: lidStatus == 'open' && heatingMode!='boiling'
```

The pre-condition is saying that; any service that executes the method `Electric Kettle.openLid()` must assure the kettle is not in the boiling status before executing the method. On the other hand, the post-condition prescribes that; the method must be implemented so that when completed, the lid is opened and the status does not change to boiling mode. Once the contract is broken, an exception is thrown to the HNS to conduct an appropriate action.

5.3   Describing Global Safety Properties

The global safety properties depend on the contents of each integrated service. Hence, it is reasonable to specify $GlobalProp(s)$ as DbC contracts in `Service` or its sub classes. For example, the global safety property G2 in Section 3.3 can be encoded as the following contract embedded in `CookingPreparationService`:

```
Contractor:      CookingPreparationService class
Class Invariant: GasValve.workingStatus=='open'
                        ->Ventilator.powerStatus=='ON'
```

The above contract prescribes a condition that; at any time when the gas valve is opened, the ventilator must be turned on. This contract is a class invariant, which must be guaranteed no matter what operations are executed within the integrated service.

5.4   Describing Environment Safety Properties

Since the environment safety properties are derived from residential issues, Home class should be in charge of $EnvProp(s)$. For instance, the environment safety property E1 in Section 3.4 can be encoded as follows:

```
Contractor:      Home class
Class Invariant:
    home.currentEnvironment.getTotalConsumption()<=30
```

The method getTotalConsumption() is supposed to return the current total consumption of electricity, which is computed from the appliances that are being turned on. This contract is also an invariant, which must be assured whatever services or appliances are operated.

5.5   Implementing Safety Validation by JML

If the proposed model is implemented in the Java language, we can use the *JML (Java Modeling Language)* [3,9] extensively for implementing the safety validation. The JML is a specification language that can be used to describe the DbC contracts in the form of Java comments, called *JML annotations*. The source code with the JML annotations is compiled by the JML compiler into *instrumented bytecode* implementing assertion-based checking routines of the DbC contracts.

Fig. 3 shows a JML annotation describing the contract mentioned in Section 5.2. In the figure, the contract is described as the annotation just above the method openLid(). The line starting with requires (or ensures) represents the pre-condition (or post-condition, respectively) of the contract. The word spec_public is for exporting the subsequent attribute to be used in the JML annotation.

Fig. 4 depicts the proposed framework of the safety validation. For given implementations of the appliance, service and home objects, a validator first describes safety properties to be validated in the JML annotation. Next, the annotated source codes are automatically compiled into instrumented bytecodes with the JML compiler. The validator also generates *test suites* against the HNS and the integrated services. For this, the validator develops the suites based on the logic and parameter values either manually or using test-case generation tools (e.g., TOBIAS [1,4]). Finally, the instrumented bytecodes are validated against

```
public class ElectricKettle {
  private /*@spec_public@*/ LidStatus  lid;
  private /*@spec_public@*/ HeatingMode hstate;
...
// JML Contract for openLid() [L1]
/*@ requires hstate != "boiling" ;
  @ ensures  lid = "open" && hstate != "boiling";
  @*/
  public void openLid() {
    ... // Implementation
  }
...
}
```

Fig. 3. Safety description as JML contract (L1)

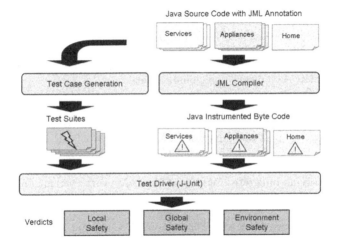

Fig. 4. Procedure of safety validation for HNS

the generated test suites by the test driver. During the test, if any JML contract is broken, the test fails. The testing can be automated using any testing framework, e.g., JUnit [10]. Thus, the safety validation for the local, global and environment properties is achieved. Note that quality and efficiency of the validation process deeply depend on the test suites. Generating *good* test suites is beyond this paper and left for our future work.

6   Conclusion

In this paper, we have formalized the concept of safety in the context of HNS integrated services. Three kinds of safety are defined: local safety, global safety and environment safety. We have then proposed an object-oriented model and safety validation framework based on the DbC with JML. The properties of the local, global and environment safety are described as DbC contracts embedded in `Appliance`, `Service` and `Home` objects, respectively. Thus, the proposed framework can assist the HNS service vendors to develop safe integrated services, systematically.

In the future work, we will conduct experimental evaluation of the safety validation against the actual HNS. We also plan to examine effective test case generation techniques in the context of the HNS integrated services. The safety validation counting the feature interaction problem is also a challenging issue for our future research.

References

1. du Bousquet, L., Ledru, Y., Maury, O., Bontron, P.: A case study in JML-based software validation. In: Proceedings of 19th Int'l. IEEE Conf. on Automated Software Engineering (ASE'04), Linz, pp. 294–297. IEEE Computer Society Press, Washington, DC (2004)
2. International Electrotechnical Commission. Household and similar electrical appliances — Safety. IEC 60335-1 (September 2006)
3. Leavens, G.T., Cheon, Y.: Design by Contract with JML. Java Modeling Language Project, Internet (2003) http://www.jmlspecs.org
4. Ledru, Y., du Bousquet, L., Maury, O., Bontron, P.: Filtering TOBIAS combinatorial test suites. In: Proceedings of ETAPS/FASE'04 - Fundamental Approaches to Software Engineering. LNCS, vol. 2984, Springer-Verlag, Heidelberg (2004)
5. Leelaprute, P., Nakamura, M., Tsuchiya, T., Matsumoto, K., Kikuno, T.: Describing and Verifying Integrated Services of Home Network Systems. In: Proc of 12th Asia-Pacific Software Engineering Conference (APSEC), pp. 549–558 (December 2005)
6. Meyer, B.: Applying Design by Contract. IEEE Computer 25(10), 40–51 (1992)
7. Nakamura, M., Tanaka, A., Igaki, H., Tamada, H., Matsumoto, K.: Adapting Legacy Home Appliances to Home Network Systems Using Web Services. In: Proc. of Int'l Conf. on Web Services (ICWS 2006), pp. 849–858 (September 2006)
8. Nakamura, M., Igaki, H., Matsumoto, K.: Feature Interactions in Integrated Services of Networked Home Appliances -An Object-Oriented Approach. In: Proc. of Int'l. Conf. on Feature Interactions in Telecommunication Networks and Distributed Systems (ICFI'05), pp. 236–251 (July 2005)
9. The Java Modeling Language - JML http://www.cs.iastate.edu/~leavens/JML/
10. JUnit, Testing Resources for Extreme Programming http://www.junit.org/

Fusion and Fission: Improved MMIA for Multi-modal HCI Based on WPS and Voice-XML

Jung-Hyun Kim and Kwang-Seok Hong

School of Information and Communication Engineering, Sungkyunkwan University, 300, Chunchun-dong, Jangan-gu, Suwon, KyungKi-do, 440-746, Korea
kjh0328 @skku.edu, kshong@skku.ac.kr
http://hci.skku.ac.kr

Abstract. This paper implements the Multi-Modal Instruction Agent (hereinafter, MMIA) including a synchronization between audio-gesture modalities, and suggests improved fusion and fission rules depending on SNNR (Signal Plus Noise to Noise Ratio) and fuzzy value, based on the embedded KSSL (Korean Standard Sign Language) recognizer using the WPS (Wearable Personal Station) and Voice-XML. Our approach fuses and recognizes the sentence and word-based instruction models that are represented by speech and KSSL, and then translates recognition result that is fissioned according to a weight decision rule into synthetic speech and visual illustration (graphical display by HMD-Head Mounted Display) in real-time. In order to insure the validity of our approach, we evaluate performance with the average recognition rates and the recognition time of MMIA. In the experimental results, the average recognition rates of the MMIA for the prescribed 65 sentential and 156 word instruction models were 94.33% and 96.85% in clean environments, and 92.29% and 92.91% were shown in noisy environments. In addition, the average recognition time is approximately 0.36 ms in given both environments.

1 Introduction

Current literature on multimodal interaction describes the potential of multimodality in terms of increased adaptability, robustness and efficiency, as well as a multi-modal HCI application takes advantage of the multi-sensory nature of humans. To name only a few examples, many original PDAs or the hand-held devices are referred to the graphical, textual, auditory information and the control sequences (such as movements of the computer mouse and selections with the touch-screen), and Multimodal systems proved to be a viable aid for visually impaired users, an alternative to WIMP (Windows, Icon, Menu, and Pointer) interfaces in mobile computing, or an entertaining extension of computer games. In addition, Grasso et al. point out the potential of combining the advantages of direct manipulation and speech interaction [1], [2]. Namely, in the desktop PC and wire communications net-based traditional computer science and HCI, according as the user interface referred to the graphical, textual, auditory information and the control sequences (such as movements of the computer mouse and selections with the touch-screen), generally they have some

T. Okadome, T. Yamazaki, and M. Mokhtari (Eds.): ICOST 2007, LNCS 4541, pp. 141–152, 2007.

restrictions and problems such as conditionality on the space, limitation of motion and so on. However, the next generation HCI for more advanced and personalized PC system such as wearable computer and PDA based on wireless network and wearable computing, may require and allow new interfaces and interaction techniques such as tactile interfaces with haptic feedback methods, and gesture interfaces based on hand gestures, or mouse gestures sketched with a computer mouse or a stylus, to serve different kinds of users. Namely, for perceptual experience and behavior to benefit from the simultaneous stimulation of multiple sensory modalities that are concerned with human's (five) senses, fusion and fission technologies of the information from these modalities are very important and positively necessary.

Consequently, we implement MMIA including synchronization between audio-gesture modalities by coupling the WPS-based embedded KSSL recognizer with a remote Voice-XML user, for improved multi-modal HCI in noisy environments. In contrast to other proposed multi-modal interaction approaches, our approach is unique in two aspects: First, because the MMIA provides different weight and a feed-back function in individual (speech or gesture) recognizer, according to SNNR and fuzzy value, it may select an optimal instruction processing interface under a given situation or noisy environment, and can allow more interactive communication functions in noisy environment. Second, according as the MMIA fuses and the sentence and word-based instruction models that are represented by speech and KSSL, and then translates recognition result, which is fissioned according to a weight decision rule into synthetic speech and graphical display by HMD-Head Mounted Display in real-time, it provides a wider range of personalized information more effectively.

This paper is organized as follows. In section 2, we describe the WPS-based embedded KSSL recognizer for ubiquitous computing. In section 3, we describe briefly web-based speech recognition and synthesis using Voice-XML. In section 4, we introduces improved the fission depending on a weight decision rule for simultaneous multi-modality, and suggests the MMIA including synchronization between audio-gesture modalities. In section 5, we evaluate and verify suggested the MMIA with experimental results for prescribed the instruction models. Finally, this study is summarized in section 6, together with an outline of challenges and future directions.

2 WPS-Based Embedded KSSL Recognizer

For the WPS-based embedded KSSL recognizer, we used 5DT company's wireless data gloves and Fastrak® which are popular input devices in the haptic application field, and utilized blue-tooth module for the wireless sensor network [3]. The structural motion information of each finger in the data glove are captured by f1=thumb, f2=index, f3=middle, f4=ring and f5=little in a regular sequence. Each flexure value has a decimal range of 0 to 255, with a low value indicating an inflexed finger, and a high value indicating a flexed finger. In addition, the Fastrak® is a solution for the position / orientation measuring requirements of applications and environments, and is a 3D digitizer and a quad receiver motion tracker, making it correct for a wide range of applications requiring high resolution, accuracy, and range. By computing the position and orientation of a small receiver as it moves through space, it provides dynamic, real-time measurements of position (X, Y, and Z Cartesian coordinates) and orientation (azimuth, elevation, and roll). Also, it provides 4ms latency updated at 120 Hz, gesture data is transmitted via RS-232 and blue-tooth module to the host at up to 115.2 K Baud, and

uses patented low-frequency magnetic transducing technology, there's no need to maintain a clear line-of-sight between receivers and transmitters. [4].

In addition, the wearable computer is a small portable computer designed to be worn on the body during use. The wearable computer differs from PDAs, which are designed for hand-held use, although the distinction can sometimes be blurry. In this paper, as the WPS (a wearable platform for the next-generation PC), the i.MX21 test board was selected, which is a next-generation PC platform in the Rep. of Korea, for application to our system. The i.MX21 test board consists of an ARM926EJ-S (16KB I-Cache, 16KB D-Cache) CPU, and includes ARM Jazelle technology for Java acceleration and MPEG-4 and H.263 encode/decode acceleration. The i.MX21 application processing block diagram is shown in Fig. 1 [5].

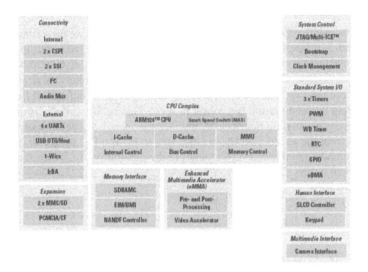

Fig. 1. The i.MX21 application processing block diagram

2.1 Feature Extraction and Recognition Models Using RDBMS

KSSL, used by the deaf community in the Rep. of Korea, is a complex visual-spatial language that uses manual communication instead of sound, to convey meaning, by simultaneously combining hand shapes, orientation and movement of arms or body, and facial expressions to fluidly express a speaker's thoughts [6]. Because KSSL is very complicated, and consists of considerable numerous gestures, and motions, it is impossible to recognize all dialog components used by the hearing-impaired. Therefore, we selected 32 basic KSSL motion gestures and 28 hand gestures connected with a travel information scenario, according to "Korean Standard Sign Language Tutor (KSSLT) [7]". KSSL motion gestures and hand gestures are classified by an arm's movement, hand shape, pitch and roll degree. Consequently, we constructed 65 sentential and 156 word instruction models by coupling KSSL hand gestures with motion gestures that are referred to KSSLT. In addition, for a clustering method to achieve efficient feature extraction and construction of training / recognition models based on distributed computing, we utilize and introduce an improved RDBMS (Relational Data-Base Management System) clustering module, which has the capability to recombine data items from different files, providing powerful tools for data usage [8], [9].

2.2 KSSL Recognition Using Max-Min Composition of Fuzzy Relation

As the fuzzy logic for KSSL recognition, we applied trapezoidal shaped membership functions for representation of fuzzy numbers-sets, and utilized the fuzzy max-min composition.

$$\text{For } (x, y) \in A \times B, \ (y, z) \in B \times C,$$

$$\mu_{S \bullet R}(x, z) = \underset{y}{Max} \ [Min \ (\mu_R(x, y), \mu_S(y, z))] \tag{1}$$

Two fuzzy relations R and S are defined in sets A, B and C (we prescribed the accuracy of hand gestures and basic KSSL gestures, object KSSL recognition models as the sets of events that occur in KSSL recognition with the sets A, B and C). That is, $R \subseteq A \times B$, $S \subseteq B \times C$. The composition $S \cdot R = SR$ of two relations R and S is expressed by the relation from A to C, and this composition is defined in Eq. (1) [10], [11]. $S \cdot R$ from this elaboration is a subset of $A \times C$. That is, $S \cdot R \subseteq A \times C$. If the relations R and S are represented by matrices M_R and M_S, the matrix $M_{S \bullet R}$ corresponding to $S \cdot R$ is obtained from the product of M_R and M_S; $M_{S \bullet R} = M_R \cdot M_S$. The matrix $M_{S \bullet R}$ represents max-min composition that reason and analyze the possibility of C when A occurs, and it is also given in Fig. 2.

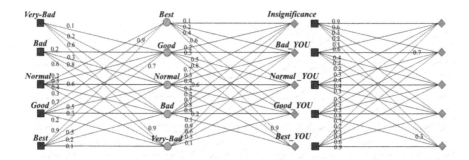

Fig. 2. Composition of fuzzy relation

WPS-based embedded KSSL recognizer calculates and produces a fuzzy value from the user's dynamic KSSL via a fuzzy reasoning and composition process, and then decides and recognize user's various KSSL according to produced fuzzy value. The flowchart of KSSL recognizer is shown in Fig. 5 (in section 4) together with an outline and flow-chart of the MMIA.

3   Speech Recognition and Synthesis Using Voice-XML

Voice-XML is the W3C's standard XML format for specifying interactive voice dialogues between a human and a computer. For ASR-engine in architecture of W3C's

VXML 2.0 [12], we used the HUVOIS solution that is Voice-XML-based voice soft-ware developed by KT Corp. in Korea for those with impaired sight that converts online text into voice and reads out the letters and words punched in through the computer keyboard, thus enabling them to use computers and the internet. The HUVOIS solution consist of HUVOIS-ARS based on HMM, TTS using tri-phone unit and HUVOIS Voice-XML, and supports client-sever network, LSS(Load Share Server) and modular structure.

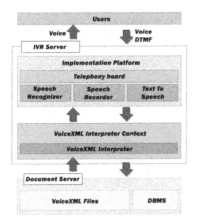

Fig. 3. The Voice-XML's architecture

The Voice-XML's architecture is shown in Fig. 3. A document server (e.g. a web server) processes requests from a client application, the Voice-XML interpreter, through the VXML interpreter context. The server produces Voice-XML documents in reply, which are processed by the Voice-XML interpreter. The Voice-XML interpreter context may monitor user inputs in parallel with the Voice-XML interpreter. For example, one Voice-XML interpreter context may always listen for a special escape phrase that takes the user to a high-level personal assistant, and another may listen for escape phrases that alter user preferences like volume or text-to-speech characteristics. The implementation platform is controlled by the Voice-XML interpreter context and by the Voice-XML interpreter.

4 Fusion and Fission Schemes Between Modalities

4.1 Fusion Technology Between Modalities

The integration scheme consists of seven major steps: 1) the user connects to Voice-XML server via PSTN and internet using telephone terminal and WPS based on wire-less networks (including middleware), and then inputs prescribed speech and KSSL, 2) the user's speech data, which are inputted into telephone terminal, is transmitted to ASR-engine in Voice-XML, then ASR results are saved to the MMDS (Multi-Modal Database Server; The MMDS is the database responsible for synchronizing data between speech and KSSL gesture), 3) user's KSSL data, which are inputted into WPS, are recognized by embedded KSSL recognizer, then the WPS transmits and saves recognition results to the MMDS, using middleware over TCP/IP protocol and wire-less networks(blue-tooth module), 4) at this point, the user's KSSL and speech data run the synchronization session using internal SQL logic of the MMDS, 5) while suggested the MMIA runs comparison arithmetic (validity check) on ASR and KSSL recognition results with pre-scribed instruction models by internal SQL logic, the NAT(Noise Analysis Tool) analyzes noise for user's speech data (wave file) which is recorded by Voice-XML, 6) According to analyzed noise and arithmetic result, the MMIA gives weight into an individual (gesture or speech) recognizer, 7) finally, user's intention is provided to the user through TTS and visualization. The suggested fusion architecture and flowchart of MMIA are shown in Fig. 4 and 5.

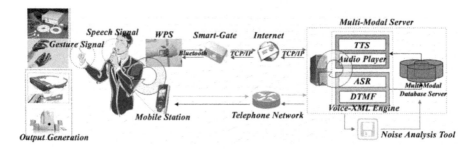

Fig. 4. The components and fusion architecture

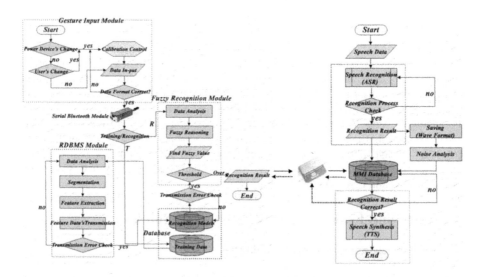

Fig. 5. The flowchart of the MMIA integrating 2 sensory channels with speech and gesture

4.2 Database-Based Synchronization Between Modalities

Human's utterance and gesture representation usually have some time difference and such asynchronous control should occur necessarily in speech synthesis processing of recognition results. All multi-modal inputs are synchronized, because while speech recognizer generates absolute times for words, gesture movements generate *{x, y, t}* triples, and initial work identifies an object or a location from gesture inputs accordingly, as speech understanding constrains gesture interpretation. This paper solves the asynchronous control problems between speech and gesture signals using a web-logic and word-unit input method based on the MMDS in section 4.1. For synchronization between speech and gesture signals, after individual speech and KSSL recognizer recognizes inputted speech and the KSSL recognition models, they transmit recognition results into the MMDS for weight application. However, the transmission time of recognition results has some time delay because of asynchronous communication of two input signals. As a result, the speech and KSSL recognition results based on word-unit are recorded sequentially to the MMDS, and while the DB is kept in standby mode via internal web-logic in case one was not input among the two input signals (where, two input signals are the recognition results of speech and KSSL), apply weights according to SNNR and fuzzy value, in the case where all input values are recorded.

4.3 Noise Analysis

In noisy environments, speech quality is severely degraded by noises from the surrounding environment and speech recognition systems fail to produce high recognition rates [13], [14]. Consequently, we designed and implemented Noise Analysis Tool (NAT) for weight decision in individual (gesture or speech) recognizer. The NAT calculates average energy (mean power; [dB]) for a speech signal that recorded by wave-format in the Voice-XML; and then computes SNNR by Eq. (2), where, *P* is average energy (mean power; [dB]), and the flowchart of NAT are shown in Fig. 6.

$$SNNR(dB) \quad = \quad 10 \log_{10} \quad \frac{P \; signal + noise}{P \; noise} \tag{2}$$

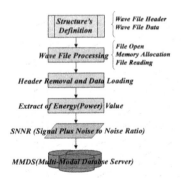

Fig. 6. The flowchart of the MMIA integrating 2 sensory channels with speech and gesture

4.4 Fusion and Fission Rules Using SNNR and Fuzzy Value

Speech recognition rate does not usually change to a SNNR of 25 dB, but if the rate lowers, the speech recognition rate falls rapidly. Therefore, the MMIA provides feed-back function according to SNNR critical value. In case SNNR critical value for weight decision is ambiguous, according as a feed-back function requests re-input (speech and KSSL) to user for clear a declaration of intention, more improved instruction processing is available. In addition, we utilized an average speech recognition rate as speech probability value for weight decision, and to define speech probability value depending on SNNR, we repeatedly achieved speech recognition experiments 10 times with the 20 test speech recognition models in noisy and clean environments, for every 5 reagents. The average speech recognition rates are given in Table 1.

Table 1. Weight values according to the SNNR and critical values for the feed-back function

SNNR Critical value	Weight value (%)		Average speech recognition rates for the 20 test recognition models (%)						
	Speech (W_S)	KSSL(W_G)	Reagent 1	Reagent 2	Reagent 3	Reagent 4	Reagent 5	Average(S)	Difference
more than 40 [dB]	99.0	1.0	98.2	98.4	97.9	98.5	98.2	98.2	0.9
35 [dB] ≤ SNNR < 40 [dB]	98.0	2.0	97.8	97.3	96.6	97.1	97.5	97.3	0.3
30 [dB] ≤ SNNR < 35 [dB]	96.0	4.0	97.5	96.5	96.6	97.0	97.4	97.0	0.2
25 [dB] ≤ SNNR < 30 [dB]	94.0	6.0	97.2	96.5	96.5	96.9	96.9	96.8	0.2
20 [dB] ≤ SNNR < 25 [dB]	92.0	8.0	96.9	95.9	96.4	96.8	96.8	96.6	2.2
15 [dB] ≤ SNNR < 20 [dB]	Feed-Back		92.4	96.2	93.8	95.2	94.1	94.3	11.1
10 [dB] ≤ SNNR < 15 [dB]	6.0	94.0	83.6	83.4	83.5	82.6	83.2	83.3	8.8
5 [dB] ≤ SNNR < 10 [dB]	4.0	96.0	71.9	72.5	70.2	79.5	75.6	74.5	22.4
0 [dB] ≤ SNNR < 5 [dB]	2.0	98.0	53.4	51.3	52.6	51.6	51.3	52.0	14.0
less than 0 [dB]	1.0	99.0	38.5	37.6	37.5	38.2	38.5	38.1	-

$$P\_W = W_S \times S + W_G \times G \tag{3}$$

- P_W : an expected value after weight application
- W_S : Defined Weight for Speech recognition mode in Table 1.
- W_G : Defined Weight for KSSL recognition mode in Table 1.
- S : speech probability (an average speech recognition rate)
- G : KSSL probability (the critical value depending on normalized fuzzy value)

$$G = \frac{Fuzzy\,Value\_Current}{Fuzzy\,Value\_Max} = \frac{Fuzzy\,Value\_Current}{3.5} \tag{4}$$

- Fuzzy Value_Current : Fuzzy value to recognize current gesture(KSSL)
- Fuzzy Value_Max = 3.5 : The maximum fuzzy value for KSSL recognition

For fusion and fission rules depending on SNNR and fuzzy value, we defined P_W that is an expected value after weight application and the *KSSL probability* (G) of the embedded KSSL recognizer in Eq. (3) and (4). This P_W value depending on SNNR and fuzzy value gives standard by which to apply weights, and because *KSSL probability* (G) is changed according to *Fuzzy Value_Current*, the P_W is changed justly. Where, the maximum fuzzy value for KSSL recognition is 3.5, and the minimum critical value is 3.2. As a result, if P_W value is over than 0.917, the MMIA fissions and returns recognition result of speech recognizer based on Voice-XML, while the MMIA fissions the embedded KSSL recognizer in case P_W value is less than 0.909. The P_W values depending on SNNR and fuzzy value are given in Table 2.

Table 2. In case *Fuzzy Value_Current* is 3.2, *P_W* values using the Eq. (3) and (4)

SNNR	Speech		KSSL		P_W
	W_S	S	W_G	G	
more than 40 [dB]	0.99	0.982	0.01	0.914	0.981
35 [dB] ≤ SNNR < 40 [dB]	0.98	0.973	0.02	0.914	0.972
30 [dB] ≤ SNNR < 35 [dB]	0.96	0.970	0.04	0.914	0.968
25 [dB] ≤ SNNR < 30 [dB]	0.94	0.968	0.06	0.914	0.965
20 [dB] ≤ SNNR < 25 [dB]	0.92	0.966	0.08	0.914	0.917
15 [dB] ≤ SNNR < 20 [dB]	Feed-Back				
10 [dB] ≤ SNNR < 15 [dB]	0.06	0.833	0.94	0.914	0.909
5 [dB] ≤ SNNR < 10 [dB]	0.04	0.745	0.96	0.914	0.907
0 [dB] ≤ SNNR < 5 [dB]	0.02	0.520	0.98	0.914	0.906
less than 0 [dB]	0.01	0.381	0.99	0.914	0.909

5 Experiments and Results

This paper combines natural language and artificial intelligence techniques to allow HCI with an intuitive mix of speech and gesture based on WPS and Voice-XML, and the experimental set-up is as follows. The distance between the KSSL input module and the WPS with a built-in KSSL recognizer approximates radius 10M's ellipse form. In KSSL gesture and speech, we move the data gloves and the motion tracker to the prescribed position. For every 15 reagents, we repeat this action 10 times in noisy and clean environments. While the user inputs KSSL using data gloves and motion tracker, and speaks using the blue-tooth headset in a telephone terminal. In experimental results, the average recognition rates of the MMIA for the prescribed 65 sentential and 156 word instruction models were 94.33% and 96.85% in clean environments, and 92.29% and 92.91% were shown in noisy environments. The uni-modal and the MMIA's average recognition rates in noisy and clean environment are shown in Table 3 and Fig. 7.

Table 3. The MMIA's average recognition results for 15 reagents

Evaluation / Reagent	Uni-modal Instruction Processing Interface						The MMIA			
	KSSL (%)		Speech (%)				KSSL + Speech (%)			
	Noise or Clean		Noise		Clean		Noise		Clean	
	sentence	word	sentence	word	sentence	word	sentence	word	sentence	word
Reagent 1	92.7	92.7	72.1	75.3	97.5	98.3	92.6	92.6	97.4	98.3
Reagent 2	92.5	93.7	70.9	78.6	94.6	95.7	92.5	93.7	94.6	95.6
Reagent 3	92.9	94.1	68.5	75.9	93.7	95.3	92.9	94.1	93.7	95.3
Reagent 4	92.8	93.1	68.9	75.9	95.1	96.7	92.8	93.1	95.1	96.6
Reagent 5	93.0	93.4	68.3	75.8	94.7	96.8	93.0	93.3	94.7	96.8
Reagent 6	92.8	94.3	68.7	76.7	93.9	95.4	92.8	94.3	93.9	95.4
Reagent 7	92.8	93.1	70.1	75.9	94.8	95.5	92.7	93.1	94.8	95.4
Reagent 8	92.5	94.6	72.1	75.3	95.3	96.3	92.5	94.6	95.3	96.3
Reagent 9	92.9	93.4	69.9	76.2	95.7	97.6	92.9	93.3	95.7	97.6
Reagent10	92.9	93.1	67.6	75.2	96.3	97.5	92.8	93.1	96.3	97.5
Reagent11	92.5	93.7	70.1	73.1	96.7	97.3	92.5	93.7	96.6	97.3
Reagent12	92.3	92.4	68.4	75.1	95.3	96.5	92.3	92.4	95.3	96.5
Reagent13	92.5	92.8	70.1	75.3	95.4	96.4	92.5	92.8	95.4	96.4
Reagent14	92.3	93.1	70.3	74.9	96.7	97.3	92.3	93.1	96.7	97.3
Reagent15	93.1	94.1	72.1	74.3	95.4	97.5	93.1	94.1	95.4	97.5
Average	92.70	93.44	69.87	75.57	95.41	96.67	92.68	93.42	95.39	96.65

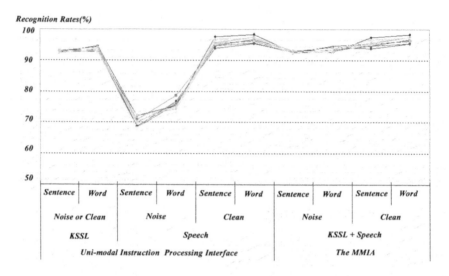

Fig. 7. The comparison chart for the MMIA's average recognition results

Table 4. The MMIA's average recognition time for 15 reagents

Evaluation / Reagent	Uni-modal Instruction Processing Interface						The MMIA			
	KSSL (ms)		Speech (ms)				KSSL + Speech (ms)			
	Noise or Clean		Noise		Clean		Noise		Clean	
	sentence	word	sentence	word	sentence	word	sentence	word	sentence	word
Reagent 1	0.33	0.26	0.24	0.22	0.23	0.21	0.39	0.35	0.34	0.32
Reagent 2	0.35	0.24	0.24	0.20	0.22	0.20	0.38	0.36	0.37	0.35
Reagent 3	0.32	0.27	0.23	0.23	0.24	0.22	0.39	0.35	0.35	0.32
Reagent 4	0.35	0.24	0.23	0.22	0.22	0.21	0.38	0.36	0.38	0.34
Reagent 5	0.31	0.25	0.24	0.21	0.22	0.20	0.40	0.37	0.36	0.34
Reagent 6	0.32	0.23	0.24	0.23	0.23	0.21	0.38	0.36	0.36	0.32
Reagent 7	0.35	0.21	0.24	0.23	0.21	0.20	0.41	0.36	0.35	0.33
Reagent 8	0.37	0.24	0.24	0.22	0.23	0.22	0.41	0.36	0.35	0.30
Reagent 9	0.34	0.25	0.22	0.21	0.21	0.20	0.42	0.37	0.36	0.34
Reagent10	0.32	0.23	0.23	0.21	0.23	0.21	0.40	0.36	0.35	0.32
Reagent11	0.33	0.27	0.22	0.20	0.22	0.20	0.39	0.36	0.34	0.32
Reagent12	0.35	0.24	0.24	0.22	0.22	0.21	0.38	0.36	0.36	0.35
Reagent13	0.37	0.26	0.23	0.21	0.22	0.21	0.39	0.37	0.34	0.32
Reagent14	0.34	0.27	0.24	0.22	0.22	0.21	0.40	0.35	0.35	0.33
Reagent15	0.36	0.25	0.24	0.21	0.21	0.20	0.40	0.36	0.37	0.32
Average	0.34	0.25	0.24	0.22	0.22	0.21	0.40	0.36	0.36	0.33

   More importantly, according as the average recognition time has exerted a strong influence on the technical capabilities of the entire recognition system, we estimated the average recognition time in the recognition processing. The experimental results are shown in Table 4 and Fig. 8 respectively. Note, these experiments were achieved under the following experimental conditions and weights.

■   *In a noisy environment, the average SNNR using actual waveform data is re-corded in laboratory space, including the music and the mechanical noise, was about 13.59[dB].In addition, the average SNNR using actual waveform data that remove noise elements for a clean environment was about 38.37[dB].*

■ *Note, if the SNNR changes due to experimental conditions such as the mu-sic and the mechanical noise, because weight has changed, then the experiment re-sult can change.*

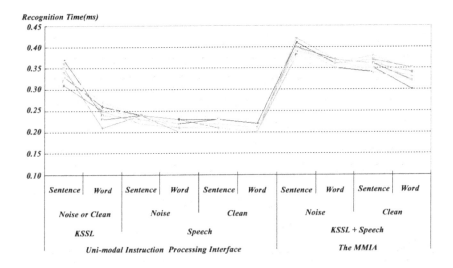

Fig. 8. The comparison chart for the MMIA's average recognition time

6   Conclusions

The advantage of multiple modalities is increased usability: the weaknesses of one modality are offset by the strengths of another. Namely, on a mobile device with a small visual interface and keypad, a word may be quite difficult to type but very easy to say. Consequently, the MMIA may select an optimal instruction processing inter-face under a given situation or noisy environment, and can allow more interactive communication functions in noisy environment, because of the weaknesses of audio-modality in noisy environments are offset by the strengths of gesture modality.

In this paper, we suggested and implemented the MMIA including synchronization between audio-gesture modalities, and the MMIA supports two major types of multi-modality, which are simultaneous multi-modality and sequential multi-modality, for instruction recognition based on embedded and ubiquitous computing.

Finally, we expect the function that understand life and culture of deaf people (and an aphasiac) and connect with modern society of hearing people through the MMIA fusing the embedded KSSL and speech recognizer.

Acknowledgement

This research was supported by MIC, Korea under ITRC IITA-2006-(C1090-0603-0046).

References

1. Grasso, M.A., Ebert, D.S., Finin, T.W.: The integrality of speech in multimodal interfaces. ACM Trans. Comput.-Hum. Interact. 5(4), 303–325 (1998)
2. Perlman, G., et al.: HCI Bibliography.: Human-Computer Interaction Resources, http://www.hcibib.org/
3. Kim, J.-H., et al.: Hand Gesture Recognition System using Fuzzy Algorithm and RDBMS for Post PC. In: Proceedings of FSKD2005. LNCS (LNAI), vol. 3614, pp. 170–175. Springer-Verlag, Berlin Heidelberg New York (2005)
4. Kim, J.-H., et al.: An Implementation of KSSL Recognizer for HCI Based on Post Wearable PC and Wireless Networks KES 2006. LNCS (LNAI), vol. 4251 Part I, pp. 788–797. Springer-Verlag, Berlin Heidelberg New York (2006)
5. i.MX21 Processor Data-sheet http://www.freescale.com/
6. Kim, S.-G.: Standardization of Signed Korean. Journal of KSSE, vol. 9. KSSE (1992)
7. Kim, S.-G.: Korean Standard Sign Language Tutor, 1st, Osung Publishing Company, Seoul (2000)
8. Oracle 10g DW Guide http://www.oracle.com
9. Kim, J.-H., Hong, K.-S.: An Implementation of the Real-Time KSSL Recognition System based on the Post wearable PC. In: Proc. ICCS 2006. Lecture Notes in Computer Science, Part IV, vol. 3994, Springer-Verlag, Berlin, Heidelberg, New York (2006)
10. Chen, C.H.: Fuzzy Logic and Neural Network Handbook, 1st edn. McGraw-Hill, New York (1992)
11. Vasantha kandaswamy, W.B.: Smaranda Fuzzy Algebra. American Research Press, Seattle (2003)
12. McGlashan, S., et al.: Voice Extensible Markup Language (VoiceXML) Version 2.0. W3C Recommendation http://www.w3.org (1992)
13. Martin, W. H.: DeciBel-The New Name for the Transmission Unit, Bell System Technical Journal (January 1929)
14. NIOSH working group.: STRESS...AT WORK NIOSH, Publication No. 99-101,U.S. National Institutes of Occupational Health (2006)

Smart Learning Buddy*

Yuan Miao

School of Computer Science and Mathematics
Footscray Park Campus, Victoria University
Ballarat Road, Footscray, PO Box 14428
Melbourne VIC 8001 Australia
Yuan.Miao@vu.edu.au

Abstract. It is common for nowadays family with two working parents. Children therefore has more unsupervised time than before. Being alone, they are more likely to avoid hard subjects, like math. They could also improperly spend their time and energy. Internet addiction is one of the examples. This research designates software agent, with a life like interface, as a learning buddy to accompany kids and school age children, guide their behaviour and attract them to what they need to learn, even it is abstract and hard, like math. The learning buddy appears as a cartoon character that live in its little master's real life, with an internal cogntive model driving its behavior. The software agent model of the learning buddy adopted a layered architecture that allows the learning buddy to have different cartoon character interfaces, cognitive models and learning content/subjects easily and independently. A pilot system has been built and four sessions of experiment have been conducted. It has very successfully attracted a great interests of school age children. Every students in the experiment has finished nearly 60 math exercises without feeling bored.

Keywords: Smart home, software agent, educational game, learning, cognitive map, DCN, children.

1 Children Need a Buddy

In today's two-parent families, it's most common for both parents to work. About 70 percent of the nation's (USA) children live in two-parent households, and in 61 percent of those households, both parents are working, according to the U.S. Bureau of Labor Statistics and the Annie E. Casey Foundation's annual Kids Count report. Of the 30 percent of kids living in single-parent households, 80 percent have working parents. (Mishell 2005; Astesano 2006) Such a status is common in many countries.

A consequence of the working families is children often have a lot of unsupervised time. Although there are programmes which offer children a safer and more productive way to spend their time after school and on weekends, they are far from enough to accommodate all the children and their time.

When school-age children are left unsupervised, they may be more likely to engage in a variety of "risky" behaviours. (Danziger and Waldfogel 2000). It is not uncommon

---

* This research is supported by the Victoria University New Research Direction Grant.

T. Okadome, T. Yamazaki, and M. Mokhtari (Eds.): ICOST 2007, LNCS 4541, pp. 153–161, 2007.
© Springer-Verlag Berlin Heidelberg 2007

that many children spend a lot of time in electronic game rooms or even on the streets. How do these have impact on their children is a worry of many parents. Another challenge is that many school age children have gained good computing knowledge. Many are better than their parents. We then have enough reasons to worry even when they are at home alone. They can install a Trojan horse in the home computer, obtain all the passwords and proceed surfing freely. Besides the access to all the unwanted content available over the Internet, addiction itself is a serious problem. (Young and Liz, 2006).

In this context, the traditional scene of a parent accompany kids in their study, reading or playing game together for mental development, has been largely challenged, given the fact that an increasing proportion of families with both parents work. Without the companion and encouragement from parents, children have more difficulties in exploring "hard" subjects. Math is one of the least favorable subjects. A survey (Texas instruments, April 5, 2006) found that "Teens Lack Math Requirements for Hottest Careers" and reported that even though four out of five teenagers believe mathematics is important for achieving their goals of being doctors, scientists, executives and lawyers, only half were planning to take advanced mathematics classes beyond their schools' minimum requirements. In another survey, (Raytheon, 2005) it was concluded that American middle school students (84%) would rather do one of the following: clean their rooms, eat their vegetables, take out the garbage or go to the dentist than sit down with their mathematics homework.

This paper presents an approach using software agent in a form of life like character (cartoon), serving as a learning buddy to accompany school age children, "guide" their behavior and attract their interests in study.

2 Smart Learning Buddy: Design Objectives

A learning buddy is to accompany kids and school age children. It needs to meet the following criteria.

-    *The learning buddy should be able to attract kids*
     There are many duties can be assigned to the learning buddy, if the little master like the companion. Therefore, being able to attract kids is a fundamental requirement. Life like cartoon characters have proven to have such a power to attract kids and school age children. In this project, puppy, birds and spirits are applied as the outlook of learning buddies. This is because some research has shown that human like virtual characters can cause users to be captious. Unlike cartoons films where the characters only live in a fixed preset story, the learning buddy lives with its little master.

-    *The learning buddy communicates with the little master verbally*
     It will be ideal if the life like learning buddy could communicate with the little master naturally. However, the voice recognition and natural language processing have yet been mature enough. For the current version, the learning buddy speaks to the little master through a natural voice text to speech engine, while takes input via a handheld device (smart phone).

- *The learning buddy has some abilities to be considered real*
  Children know that cartoons are in the TV programs or cinemas that are not real in their life. The learning buddy should however, have some ability to be perceived by the little master as "something" real in his/her life. For example, the learning buddy should know the little master's name and address him/her with the name. More profound knowledge of the little master is challenging but extremely helpful. An internal cognitive model of the learning buddy would be desirable, which will be covered in the following criterion.

- *The learning buddy has an internal cognitive model to drive its behavior*
  Every person carries maps of the external world in their mind. This cognition and the corresponding inference drive the behavior of the person. Such a process differs a "creature", like a human being, from a "thing", like a television. Once the learning buddy has an internal cognitive model to drive its behavior, it is more likely to be perceived (by the little master) as someone really life in his/her world.

- *The learning content can be easily modified, for example, by the child's parents or teachers*
  There are some existing work applying life like characters in assisting learning process. (Lester 1997, Rickel 2002) However, the learning content is fixed within the system, or the character is fixed within the system, or both. This research removes this restriction and separates the life like interface, cognitive model and content to be learned. Such architecture allows the flexible update to the interface, cognitive model and content independently.

3   Smart Learning Buddy in a Smart Home

A smart learning buddy may simply reside in a computer (PC or pocket PC). However, it is more attractive if it has an environment with richer support. Smart home is such an environment can accommodate smart learning buddies.

3.1 Smart Home

Figure 3.1.1 is the diagram of a smart home environment. This is an environment providing wireless communication, indoor positioning system, pervasive computing infrastructure and importantly, mobile agent platform. It is a general environment supporting a range of higher level services. Besides learning support to school age children, it has also been applied in learning support to pre-school kids and aged care.
   In the smart home diagram, there are six essential components:

- Server: The server is the main computing facility as the central information processing and storage unit.
- Gateway: The Gateway bridges the smart home and the external world, including the Internet and external nets like friends' smart homes, teachers and etc. In the pilot study, the Server and Gateway are physically within a same computing Server.

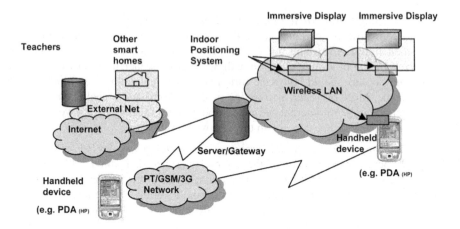

Fig. 3.1.1. Smart home environment

- Wireless LAN: It is an infrastructure facilitating the communication of different components of the smart home. In the pilot study, 802.11g wireless LAN is adopted.

- Immersive Displays: Immersive displays are visual environments where smart learning buddy live. An immersive display can have its own computing unit, or be connected to the server. In the pilot study, the immersive displays are projected walls. An alternation is using wall mounted displays like plasma screens or LCDs.

- Indoor Positioning System: The indoor positioning system tracks where the children locate. In the pilot study, middle ranged RFID reader is used at the entrance of each door. The child carries an RFID to identify himself/herself. A finer indoor positioning system can be organized if more accurate positioning information is needed. There are two widely used technologies. One is through the image processing obtained by video cameras. Another is the hybrid wave based positioning systems, using radio and ultrasound for positioning.

- Mobile Platform/Handheld Device: When the little master is away from home, or somewhere does not have an immersive display (e.g., laundry, or garden), he/she would be reminded to carry his/her handheld device. In the pilot study, a smart phone running Windows mobile 5 is used. The smart learning buddy will then "jump" onto the handheld device to follow the little master when he/she leaves the room.

Note that all the computing systems (e.g., the server, the computing units of immersive displays) should run an agent platform so that the smart learning buddy can reside).

3.2 Software Agent Model of Smart Learning Buddy

A software agent is a software entity that differs from others by being autonomous. Before the concept of software agent was proposed, computing entities are basically reactive to users' input. A software agent however, takes the active role and behaves proactively. This requires agents be goal oriented (Shen et.al 2004). An agent has its own thread of control to decide its next action/behavior. It is normally referred as the agents' internal mental activities. There are several agents internal models proposed up to date. This research applies models in cognitive map family. (C.Y. Miao et al, 2001, 2002; Y. Miao et al, 2001). Figure 3.2.1 is the diagram of the agent model of the smart learning buddy.

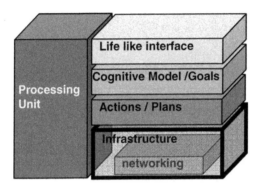

Fig. 3.2.1. Agent model for smart learning buddies

In the agent model, the processing unit provides the information processing capability. The infrastructure provides fundamental supports including networking supports. Cognitive model will be covered in Section 3.3.

An action $\mathscr{A}$ is an action that the agent could perform. It can be an animation that the agent to show, like Animation ("Suggest"); or speak something, like SAY ("Well done John!"). Actions can include parameters. It is a high level model to describe agents' behavior.

Fig. 3.2.2. Animation ("Suggest") of Peedy, Animation ("read"), and SAY ("It's time for mental training!") of Cami [1]

Actions can be organized into a plan $\mathscr{P}$. A plan has an objective, which is something the agent means to achieve via a sequence of actions. An example plan is shown in Fig. 3.2.3. Branches, loops and nested plans are allowed for complex plans.

---

[1] Peedy is originally a character of Microsoft. Cami is originally a character of Stegami.

```
SAY ("What is the sum of 21 and 12?")
Display("1) 35;\n 2)    33;\n 3) No of 1) and 2)")
Input (Answer)
If (Answer = 2 )
    {
            Animation ("Congratulate")
            SAY ("Well done John!")}
Else{
            Animation ("Blink")
            SAY ("Oh. Let's see: 2+1 is 3. 1+2 is also 3. So it is 33.") }
```

Fig. 3.2.3. A plan is a sequence of actions

Life like interface: There are many ways to achieve a life like interface. In this project, Microsoft agent (Microsoft 2003) is applied with stationary systems (e.g. a PC) and a modified version (developed locally) is used for mobile platforms.

Mobility: The agent/smart learning buddy may migrate from one system to another system. Visibly, it could jump from the handheld device onto a wall mounted display. All the systems that could accommodate the agent include an agent platform. For details on how mobile agent migrates, refer to Autran and Li (2004). The additional visualization of this migration with smart learning buddy is that the migration is always associated with an animation of hide (Animation ("hide")) at the original platform and an animation of show (Animation("shown")) at the new platform.

3.3 Cognitive Model of Smart Learning Buddy

Cognitive map is a family of models for capturing human's perception of the external world and facilitate the inference over it (Miao et al. 2006). It can be formalized by a digraph in which vertices represent concepts and arcs between the vertices indicate causal relationships between the concepts. The model has a matrix form to facilitate inference through matrix calculation or similar mathematic processes.

In this research, fuzzy cognitive map and dynamic causal net are applied as the cognitive model of the smart learning buddy. In the following part of this subsection, a few examples are used to illustrate the cognitive model of the smart learning buddy.

Figure 3.3.1 a) shows a basic cognitive map (fragment) modeling the intention of play. In a pilot study, many children would play with the virtual character for a long time and forget that they need to do math exercises. It is a common problem that many kids become addiction to computer games.

Figure 3.3.1 b) shows a modified cognitive map model (fragment). In this model, the buddy's energy is added, which prevents the buddy to play with the little master too long. Before that, it can proactively suggestion some alternations to the little master, like "I am tired, not feeling playing on."; or "I am hungry." ; or "Shall we do some math exercise?"

Figure 3.3.1 c) shows a further modification to the cognitive model fragment. It is known that children are not able to hold interests on a same thing as long as adults. The learning buddy brings that in its model and shifts subjects before the little master get bored: the buddy becomes bored before the little master becomes bored.

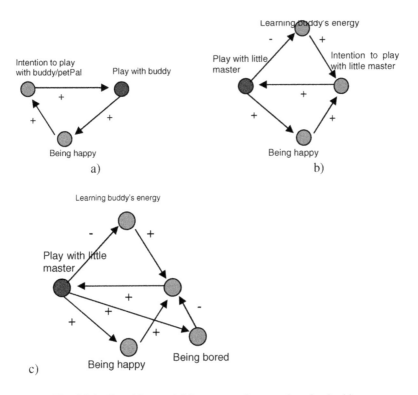

Fig. 3.3.1. Cognitive model fragments of a smart learning buddy

The cognitive model differentiates the levels of concepts, like how serious is the buddy being bored. It is classified as a mutli-value FCM or simplified DCN (Y. Miao et al. 2001, 2006).

4 Experiment Results

To verify the ability of smart learning buddy in attracting school age children, a pilot system, Math4Kids is built[2]. In this system, a puppy is used as the learning buddy to accompany children aged 7-10 years old in math study. Nine students joined the four sessions experiment and the result is extremely encouraging. A kid has even

---

[2] The cognitive model has not been implemented but was implied in the pilot system. The experiment results reported here was from the pilot study. More studies and data collection will follow, which will include agents equipped with cognitive models, collaborative learning, teachers/parents interface and comparative studies.

attempted over 140 math exercises, with the companion of the learning buddy. It would be unbelievable in normal situations. Even the least number of attempts is nearly 60, which is far over expectation in normal situations. Figure 4.1 shows a feedback from kids on the learning buddy. The feedback asked children to grade their learning buddy in five aspects, with a scale of 6 (6 is the best). Almost all kids gave their buddy the best feedback. It is better than expected.

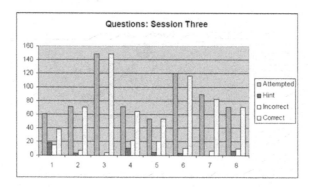

Session Four: Questionnaire Results

	Liked Petpal 4 Math	Liked Agent	Agent was friendly	Agent was helpful	Agent was smart
1	6	6	6	6	6
2	5	5	5	5	5
3	6	6	6	6	6
4	6	6	6	6	5
5	6	6	6	6	6
6	6	6	6	6	6
7	6	6	6	6	6
8	6	6	6	6	6
9	6	6	6	6	5
Mean	5.9	5.9	5.9	5.9	5.7
Standard Deviation	0.33	0.33	0.33	0.33	0.50

Fig. 4.1. Pilot experiment on the ability of smart learning buddy in attracting school age children in math study. a) Exercises attempted, number of correct and incorrect answers b) children's feedback to their learning buddy.

Acknowledgement

Sincere thanks to Dr. Xuehong Tao for her initial discussions of applying virtual characters in kids' education and the contribution to the ideas lead to the pet based educational game. These contributions directly led to the pilot project to verify the ability of virtual characters in attracting school age children. Thanks to Miss April

Perkins, who programmed the experiment system, Math4Kids and conducted the experiment in Sunshine East Primary School, as her honor project with Victoria University.

References

1. Autran, G., Li, X.N.: A practical approach to agent migration protocol. IEEE/WIC/ACM International Conference on Intelligent Agent Technology, pp. 337–340 (2004)
2. Bennett-Astesano, S.: Forget Balance, Today's Parents Are Mixing Work and Family. The Working Family in 2006 Last visited (January 2007) http://www.parenthood.com/articles.html?article_id=9408
3. Danziger, S., Waldfogel J.: Investing in Children: What Do We Know? What Should We Do? The ESRC Research Centre for Analysis of Social Exclusion (2000)
4. Kinsella, K., Phillips, D.R.: Global Aging: The Challenge of Success. Population Bulletin 60(1) (2005)
5. Lester, J.C. et al.: The persona effect: affective impact of animated pedagogical agents. the SIGCHI conference on Human factors in computing systems, pp. 359–366 (1997)
6. Miao, C.Y., et al.: A Dynamic Inference Model for Intelligent Agent. International Journal of Software Engineering and Knowledge Engineering 11(5), 509–528 (2001)
7. Miao, C.Y., et al.: Agent that Models, Reasons and Makes Decisions. Knowledge Based Systems. 15(3), 203–211 (2002)
8. Miao, Y. et al.: The Equivalence of Cognitive Map, Fuzzy Cognitive Map and Multi Value Fuzzy Cognitive Map. IEEE International Conference on Fuzzy Systems, pp. 1872–1878 (2006)
9. Microsoft Agent. last visited (January 2007) http://www.microsoft.com/msagent/default.asp
10. Mishell, L., Bernstein, J., Allegretto, S.: The State of Working America 2004-2005. ILR Press (2005)
11. Raytheon: MathMovesU Survey Highlights (2005) last visited (January 2007) https://www.mathmovesu.com/BTS/research.htm
12. Rickel, J., et al.: Toward a new generation of virtual humans for interactive experiences. IEEE Intelligent Systems and Their Applications, pp. 32–38 (2002)
13. Shen, Z.Q., et al.: Goal autonomous agent architecture. 28th Annual International Conference on Computer Software and Applications. 2, 45–46 (2004)
14. Texas Instruments: National Math Month Survey Finds Teens Lack Math Requirements for Hottest Careers (2006) Last visited (Janaury 2007) http://education.ti.com/educationportal/sites/US/nonProductSingle/about_press_release_news78.html
15. Miao, Y., Liu, Z.Q., Siew, C.K., Miao, C.Y.: Dynamic Cognitive Net-an Extension of Fuzzy Cognitive Map. IEEE Transaction on Fuzzy Systems 9(5), 760–770 (2001)
16. Young, S.C., Liz, R.: Policy and Power: The Impact of the Internet on the Younger Generation in South Korea. Social Policy and Society,Cambridge University Press, vol. 5, pp. 421–429 (2006)

Personalized Magic Mirror: Interactive Mirror Based on User Behavior

Dongwook Lee, Jieun Park, Moonheon Lee, and Minsoo Hahn

Information and Communications University, Digital Media Lab,
517-10, Dogok-dong, Kangnam-gu, Seoul, South Korea
{aalee,rashrune,monhoney,mshahn}@icu.ac.kr

Abstract. Ubiquitous computing is changing our daily life with evolving devices. A mirror, which is one of our daily tools, is also showing its role changing to the instrument of information displayer. These evolving devices already have their own traditional roles and interaction methods. Therefore, one of the most important factors for the implementation of these devices is the interaction methods between devices and a user. Based on this point of view, this paper proposes an active digital mirror system which considers the relative position and interaction between a user and mirror. To implement our goal, we studied users' behaviors and interactions on mirrors with user testing. There were four considered factors: the distance between a user and mirror, relative position of a user in front of a mirror, intuitive interaction method and personalized information. Based on the preliminary research, we designed the user interface of the mirror, and developed a prototype which has three recognition modules: a distance measuring module using infrared sensor arrays, a user recognition module by computer vision technique, and a control perception module using infrared sensor grid. In addition, the next steps for improving the user-centered digital mirror system, and the possibility for developing a mirror-shaped computer system were suggested.

Keywords: Ubiquitous Computing, Interactive Mirror, Home Automation.

1 Introduction

New paradigm named 'Ubiquitous Computing' and technological advancement of engineering are changing and improving the way and quality of modern people's life. One of the ways to change people's life is to evolve the things of daily necessity. Actually, many of daily devices had been changed and improving their roles.

This paper focused on the information displayer which is based on a mirror, one of the oldest daily devices of human beings. Digital mirror, evolved form of the traditional mirror, has not only traditional role of mirror which reflect the image of object, but also the role of information displayer. The changed role of mirror needs additional interaction methods, and this leads the necessity for the definition between a user and mirror.

In this paper, we categorized the interaction between a user and mirror based on the relative position of a user, and defined more intuitive interaction based on human

T. Okadome, T. Yamazaki, and M. Mokhtari (Eds.): ICOST 2007, LNCS 4541, pp. 162–169, 2007.
© Springer-Verlag Berlin Heidelberg 2007

behavior. The study about the interaction between a user and mirror consists of several human behavior factors: the distance between a user and mirror, relative position of a user in front of a mirror, more intuitive interaction method and personalized information.

Based on the result of the preliminary study, we implemented prototype mirror named with 'Personalized Magic Mirror' (PMM). The functions of PMM which are user recognition, user position detection, user input detection and personalized information, were implemented to provide more intuitive interaction between a user and mirror. The implemented prototype, PMM, was displayed at the exhibition hall of a conference. The result of exhibition shows that the interaction based on a human behavior was effective.

The rest of this paper consists of four sections. In section 2, the approaches and researches about digital mirror will be shown. The study about the behavior of human and development process of PMM will be shown from section 3 to section 4. Exhibition and the result will be discussed at section 5. Finally, we will conclude this paper with section 6.

2 Related Work

Originally, a mirror has the characteristics of displayer. For this reason, there were more tries to provide information through mirror surface than other devices, and there exist many researches and approaches about mirror display.

The approaches about mirror like display can be divided by two categories: mimicking the characteristic of a mirror which is to reflect the image of an object and augmenting information displayer to the mirror.

Normally, approaches to imitate the characteristics of a mirror are tend to more experimental. Most of them use a camera to take a picture of humans or objects, and display the images through the display. In case of Alexandere et al. [4], they recognized and tracked human face with camera, and then, redisplay the processed image to a user for the purpose of entertainment. For more practical approaches, there are tries to aid a driver by providing additional information. Toru et al. and Pardhy et al. [2][6] proposed the virtual mirror system which provides the visual information of driver's blind area.

One of the problems of these approaches is that it is difficult to acquire the front image of mirror like surface which is displayer. If low resolution and precision of acquired image can be acceptable, it can be possible to implement abstraction of characteristic of mirror [9]. However, this low resolution approach is not so attractive for information display. To solve this problem, Alexandre et al. [1] suggested the virtual mirror system which can acquire front image of displayer by using magnetic sensors and 3D geometric computation.

On the other hand, attempts to provide information to a user by using the surface of mirror are tend to be more human-centered, and some of them are on sale in the market. Philips Research and Miragraphy [7][8] are the cases of commercialization for the display using mirror. The display was attached to mirrors with various sizes for the information display. Fujinami et al. [5] implemented digital mirror which can provide personal information. They applied human's daily device to recognize user's identification. There is another suggested concept which use mirror display as a communication device. Panos et al. [3] suggested intra-home communication device based on the mirror display and camera. Studies on the human factors using mirror display [10] also can be found.

Differ from other approaches, this paper focused on large-scale mirror display which placed at public space. The distance between a user and mirror and relative position were studied based on the human's behavior. To develop more intuitive interaction, we considered the traditional role of mirror which is to reflect the image of objects.

3 Method

The research procedure of this paper consists of two parts: Designing User Interface (UI) based on the interaction between a user and mirror and Implementing the proto-type mirror based on preliminary research.

The UI design procedure includes four major processes: Base Research, User Test, Result Analysis and UI and Scenario Design. We defined the characteristics and functions of the digital mirror based on the UI design. The implementation procedure is based on the result of the UI design. The procedure consists of designing the system and hardware of prototype and implementing it.

3.1 UI Design Procedures

The purpose of UI design procedure is to analyze and study the interaction between a user and mirror based on the user's behavior and needs. To do this, we performed user investigation and experiments. Figure 1. briefly shows the procedure of UI design.

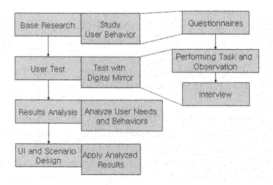

Fig. 1. Processes of User Interface Design

Base Research. Firstly, we studied the behavior of human using a mirror. Especially, we focused on the mirrors positioned in the public place like subway station or hall of the building. In these environment, we can find that the user were changing the distance between them and mirror according to their interests. For example, if they want to look at their face, they got close to the mirror. On the other hand, when the users tried to check their silhouette or their whole body, they stepped back from mirror. And we performed some questionnaires about digital mirror to users. The users' personal information like right-handed or left-handed and preferences were collected.

User Test. The user tests were performed with digital mirror system for test. This digital mirror consists of translucent half-silvered mirror and 23" LCD monitor, and can be controlled by remote control. It can be used as traditional mirror when the displayer turns off. There were 20 subjects, 10 of them were men and 10 of them were women, who are familiar with digital appliance. The subjects were asked to perform the tasks in front of the digital mirror for the test, observed during tasking, and interviewed after tasking.

The tasks which were performed by users contain two different aspects: the characteristics of traditional mirror and information displayer. The subjects were asked to check out their shape using the mirror which displays the visual information. During the tasks, the subjects tried to change the shape, position and size of the information window to find the optimal way of information displaying. This procedure was observed by four cameras which were positioned near the mirror.

The user tests were processed with two major concerns. First, because the distance between a user and mirror altered the image of mirror surface, we focused on to find where the users' interests based on the distance between a user and mirror. Secondly, we tried to analyze how the users feel and recognize the position, shape, size and necessity of the information display window. Furthermore, the interaction methods of users were observed carefully.

Result Analysis. We analyzed the result of user tests and arranged five human behaviors. First, the subjects showed more active behavior when they were a short distance away from a mirror. Second, if the subjects were a long distance away form a mirror, they were more interested in the information itself then to modifying the information window. Third, when the subjects use the mirror for its traditional characteristic, they want to put the information window away. Fourth, when the subjects use the mirror for the information displayer, they want to put the information window near their eyes. Fifth, the subjects showed very intuitive motions to interact with digital mirror and they were all non-touch style.

Fig. 2. Examples of Input Interactions

Figure 2. shows the motions of subjects for positioning information window. As the figure shows, most of the subjects preferred more intuitive and simple interaction method: the linear movement of hand or arm.

UI and Scenario Design. Based on the results of user test, we defined several interaction factors. First, the function of information display of digital mirror should not disturb the function of traditional mirror. Second, all users need their own personalized information. Third, the shape, contents and size of the information should be changed according to the distance between a user and mirror. Fourth, the relative position of a user in the range of mirror should be considered to place the information window. Fifth, the interaction between a user and mirror should be intuitive and non-touch style. Sixth, users need their picture from another point of view. Seventh, users want to place the information window near their eye position.

These interaction factors were applied as situational and functional requirements. Based on the third factor, the size for information window of digital mirror should be varies. If a user gets close to a mirror the window's size must get larger, while the user gets away from the mirror the window's size should get smaller. The fourth factor made us to adopt the relative position of a user who is in front of the mirror. Because we assumed that the mirror is placed in the public space, the user direction of walking should be considered. To avoid break the fifth factor, the walking direction of a user makes a cross to the direction of information window. Figure 3. shows the motion and size of the information window according to the user's walking direction. This means that if a user is moving from left to right, the information window should start moving from right to left. This behavior makes the user and information window meet at the center of the window. Therefore, the user can use the traditional function of mirror at anytime except at the center of the mirror. At the center of the digital mirror, the user may see the information window more precisely. If the user doesn't want to watch the information window, the window should be moved to aside. According to the fifth factor, this interaction between a user and mirror should be a simple gesture which is similar to what we do to put a thing away in the real world.

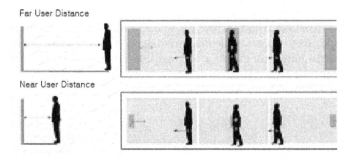

Fig. 3. User's Walking Direction and Information Window

3.2 Implementation of Personalized Magic Mirror

Based on the preliminary research about UI design and UI requirements, we implemented new digital mirror. As the procedure to implement PMM, we first designed state of PMM based on the user's input and relative position. According to the state, the whole structure of PMM was constructed with hardware and software parts.

There are three states that describe the behavior of PMM. Each state has its own role and transit to another state by the input or motion of a user. Figure 4 shows the functions and transitions of states.

The top-most state of Figure 4 is Far-Range User state. PMM in this state means that there is no user near the mirror. If a user gets close to PMM, the state transit to the second one: Near-Range User state. At this state, PMM tries to recognize a user and provide personalized information in abstract form. If a user gets more close to PMM, the state transit to the Controllable state. In this state, PMM provides personalized information in detail, and a user can interact with PMM to put away information window.

The structure of PMM is consists of four major modules: System Controller, Rendering Manager, User Manager and Private Information Manager. The whole system is managed by System Controller. The input from a user is sent to System Controller through User Manager. With this information, System Controller queries personalized to Private Information Manager for the personalized information. The personalized information is sent to Rendering Manager whose role is to create information window, and this information window was provided to a user through Personalized Magic Mirror front-end.

User Manager has three main functions. First one is user recognition. Because of the characteristics of a mirror, we adopted vision based facial recognition. By using webcam, User Manager finds and extracts the face of the user by adopting Viola's AdaBoost [11]. The extracted face features were compared with feature maps to find valid user by using HMM. Each Feature map of valid user were registered with HMM in advance.

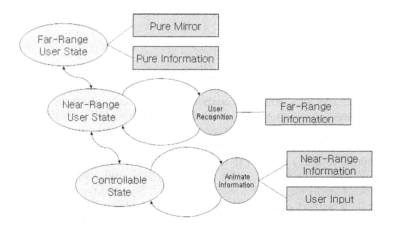

Fig. 4. States of PMM and their transitions and functions

Other functions are to find user position and user input. These functions were implemented by using Infra-Red (IR) sensors. The position of a user was detected with IR sensor array which is placed on the front of PMM. IR sensor array consists of 13

Fig. 5. Interaction between user and PMM

IR sensors. With this IR sensor array, User Manager categorizes user's position in five spots and find how far the user is. When PMM is in Controllable state, a user can interact with PMM with gesture motion. This gesture was detected by rectangle-shaped IR sensor array which is placed on the front-end of mirror surface. Rectangle-shaped means that IR sensor arrays are sensing inside of the rectangle. 8 IR sensors are placed on each the top and bottom sides, 5 IR sensors are placed on each left and right sides. To avoid the interference between IR sensors, every sensor emits IR light with duration of 100 microseconds and has 10cm distance each other. With these IR sensor arrays, User Manager can detect user's hand with 14 by 9 resolutions.

The brief description of user interaction can be shown in Figure 5. If the user moves his hand just like to put the information window away, rectangle-shaped IR sensor array detects his or her hand serially.

Rendering Manager renders the information window which will be present to a user with personalized information from Private Information Manager. The information window is rendered in 3D surface for more dynamic representation.

4   Results

Because most of the requirements from UI design are concerned with the interaction between a user and mirror, the most important point among the implementation procedure was User Manager which perceives and detects user's information.

In case of user's position detection, the IR sensor shows very sensitive behavior. Especially, if a user moves fast, the IR sensor array returns error results. However, the user's input detection shows higher performance results. It is the best way to adopt face recognition to find valid user in the circumstance of mirror based system. However, face recognition is very sensitive to the light. The initial version of PMM also showed lack of performance with user recognition. By modifying the algorithm of face recognition method, current version of PMM shows better performance.

PMM was presented in several conference exhibitions. Almost all of visitors were agreed with the user's needs and interaction method. Especially, the way to preserve function of traditional mirror was noticeable.

5   Discussion and Conclusion

In this paper, we proposed a digital mirror which has characteristics and functions based on the interactions between a user and mirror. To do this, we performed preliminary research about the behaviors of users and we made UI guideline for the digital mirror which can provide personalized information to a user. With the result of the preliminary research, we implemented a brand new digital mirror: Personalized Magic Mirror.

The appliances which were evolved by ubiquitous computing should be designed based on the user's behavior. Otherwise, the embedded functionality can disturb its main functionality and decrease user's performance. Our next research plan is to build integrated mirror which is placed in the home environment and interact with other home appliances.

Acknowledgment. This research was supported by the MIC (Ministry of Information and Communication), Korea, under the Digital Media Lab. support program supervised by the IITA (Institute of Information Technology Assessment).

References

1. Francois, A., Kang, E., Malesci, U.: A Handheld Virtual Mirror, ACM SIGGRAPH Conference Abstracts and Applications proceedings (2002)
2. Pardhy, S., Shankwitz, C.: Donath, Intelligent Vehicles Symposium, A Virtual Mirror For Assisting Drivers (2000)
3. Markopoulos, P. et al.: Personal and Ubiquitous Computing, The PhotoMirror appliance: affective awareness in the hallway (2006)
4. Darrell, T., Gordon, G., Woodfill, J., Harville, M.: A Virtual Mirror Interface using Real-time Robust Face Tracking. In: Proceedings of the Third International Conference on Face and Gesture Recognition, IEEE Computer Society Press, Washington, DC (1998)
5. Kaori, F., et al.: AwareMirror: A Personalized Display Using a Mirror, Pervasive Computing (2005)
6. Miyamoto, T., Kitahara, I., Kameda, Y., Ohta, Y.: Floating Virtual Mirrors: Visualization of the Scene Behind a Vehicle, 16th International Conference on Artificial Reality and Telexistence (2006)
7. Philips Research: HomeLab http://www.research.philips.com/technologies/misc/homelab
8. Miragraphy http://www.hitachi.co.jp/Prod/elv/jp/tosi/solution/c_tosi_solu_mirror.html
9. Rozin, D.: Wooden Mirror http://www.smoothware.com/danny/woodenmirror.html
10. Soh, Y., Hahm, W., Choi, H.: Digital Mirror User Interface, ACM SIGGRAPH 2006 Research posters SIGGRAPH '06 (2006)
11. Viola, P., Jones, M.: Asymmetric AdaBoost and a detector cascade. In: Proc. Neural Information Processing Systems (NIPS), pp. 1311–1318 (2001)

Service Abstraction Model for Dynamic User Interface Presentation

Mossaab Hariz, Mahmoud Ghorbel, and Mounir Mokhtari

Institute Nationale des Télécommunications, 9 rue Charles Fourier 91011 Evry, France
{mossaab.hariz,mahmoud.ghorbel,mounir.mokhtari}@int-evry.fr

Abstract. This paper describes a novel approach based on the user interface presentation of unknown services within a pervasive environment dedicated to dependant people. The idea is to map the heterogeneity of the environment objects into abstracted service. At the level of the user interface, this will allow the personalization and the adaptation based on user's requirements. Service provision has to be effective and performed according to current context particularly when focusing on people with special needs. The proposed Service Model (SM) supports the diversity of existing pervasive services and their content, the personalization and adaptation to both user and device profile, and the context-aware discovery in visited environment. The SM is composed of three main parts: Service Profile describing what a service can do, Service Interface describing how a service should be used and Service Logic Code dealing program code and internal functionalities of the service.

1 Introduction

Pervasive environment is growing and we have rapidly moving from conceptual systems to real products. However, the design of such environment is challenged by the heterogeneity of the available services (e.g. home control, remote alarm, message reminders, etc...). To overcome this heterogeneity, the idea is to propose a service description language based on service abstraction model. As illustrated in the figure below, the user interface (UI), which could be processed on a mobile device should be designed based on the user's requirement. In order to present the suitable services, the UI handles the abstracted services to allow user accessibility and assistance. The abstraction level depends on the definition of the service, its content, and its associated functionalities (properties).

The main challenge in assistive environment, smart space for dependant people, is in one hand, to interact with unknown services (online environment discovery), and on the other hand the personalization of this interaction according to user profile. In fact the user would like to interact with his environment dynamically without having the constraint of preliminarily services configuration. Especially, when dealing with unknown environments, such as train station, Airport, work place, etc.

Real time discovering of services in multiple living environments impose to personalize the interaction not only based on user profile, but also to contextual situation (context awareness).

T. Okadome, T. Yamazaki, and M. Mokhtari (Eds.): ICOST 2007, LNCS 4541, pp. 170–178, 2007.

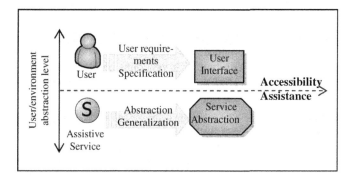

Fig. 1. From User need specification to Service abstraction

One could propose that Web Services tehnologie could solve the service discovery issue according to users requirements, but the problem is that web services describes each service with it's corresponding proprietary user interface, such as the Web is working today. This implies that two different services, such as Google and Yahoo search engines, use two different rigid graphical user interfaces (GUI) even if they have functional similarities. Merging both GUIs should decrease the complexity of the interaction expecially when dealing with increased numbers of services within smart environments for people having limited numbers of degrees of freedom (people with physical disabilities).

1.1 Smart Environment for Dependant People

Usually our living environment becomes less and less accessible for people having severe physical limitation (people with disabilities and elderly people). Smart environment could compensate the disability by providing suitable home automation (ex. open a door, TV control, etc.) and services (ex. Communication, leisure, adapted information, etc.) to the end-users.

Development of smart home technologies dedicated to people with disabilities provides a challenge in determining accurate requirements and needs in dynamic situations. User interfaces have to be adapted and personalized according to the dynamic situation of the environment.

Based on our former experimentation involving end-users, we have found out that the crucial problem in located at the level of the human machine interface which is the only way to interact with the complexity of the environment. Thus it's getting urgent to facilitate this interaction by developing flexible user interface able to be personalized.

1.2 User Interface Personalization

Selecting the most adapted input device is the first step for any system accessibility and the objective is to allow the adaptation of available functionalities according to user needs. For this purpose we have developed a software configuration tools, called ECS (Environment Configuration System), which allows a non expert to configure easily any selected input device with the help of different menus containing activities associated to environmental commands of any system.

According to human requirements, and to the selected input devices, the ECS offers the mean to associate graphically the selected actions to the input device events (buttons, joystick movements, etc.). The ECS software is actually running and fully compatible with most home equipment. It generates an XML object as standard output which will be easily downloaded in our control system (PDA)[1].

The weakness of the ECS, even if it's allow user interface personalization, is that it handles only pre-defined devices and services belonging to known environments (ex. home environment). In our case, we have to consider multiple living environments, which could not be limited to home environment, such as, the friend house, hospital, train station, etc.

This implied to consider service discovery mechanism, based on a Service Abstraction Model, allowing the dynamicity of user interface.

The contribution of this paper is around two folds: survey of existing methodologies to design a dynamic user interface and the proposition of a service abstraction model. The rest of this paper will be organized as following: First we will analyze the state of the work in the light of UI requirements, after that we propose our solution based service abstract model. In section four we will present a general view of our architecture and section five present our general demonstrator. Finally, we summarize and conclude the paper.

2 Designing a Dynamic User Interface

The interaction with pervasive environment represents a difficult task mainly when target people have special needs. Indeed, the interaction should be done through a user-friendly interface making it easier to access to the environment and benefit from assistance [2]. Consequently, the UI should support not only user preferences but also user capabilities. At the same time, several mobile terminals are proposed to deal with proposed services (PDA, Smart phone, Tablet PC); that involves another aspect of heterogeneity.

The design of dynamic user interface represents an open issue we have to consider especially when focusing on mobile devices. In that sense, several solutions were proposed in the literature in the UI design. UI design could be based on three main approaches which will be detailed below.

2.1 Application Driven Approach

In such kind of approach, the policy is to have a specific user interface of one application per target device. That could be done following two manners:

- Manually: where the same application must be developed beforehand according to each targeted device. On the run time, there will be a selection for the appropriate version based on the chosen device. In [3], authors perform the service discovery based on several parameters, among them the target device, in order to select the appropriate User Interface.

- Automatically; the user interface is generated automatically according to the requester device. In [4], authors propose a framework able to generate the GUI for different kinds of terminals in order to support adaptable graphical interfaces.

Such approach has the advantage to be easy to implement especially because the difficult work is made at the level of a UI server and not on the device itself. Nevertheless, such systems are hard to adapt to the user requirement (preference, capabilities…) and the context of use. Besides, a simple modification of the context or a new service discovered involves the reconstitution of another application including the current modification if it's possible.

2.2 Web Browser-Based Approach

In the web based approach, it is assumed that every device implement necessarily at least one web browser [5]. As a result we can develop a web based application to handle the heterogeneity of the devices. Many user interface language was built up to describe the look and presentation of the interface. Among them we can list:

2.2.1 XUL
The primary interface language of Mozilla Foundation products is XUL [6]. XUL documents are rendered by the Gecko engine, which also renders XHTML documents. It cooperates with many existing standards and technologies, including CSS, JavaScript, DTD and RDF, which makes it relatively easy to learn for people with background of web programming and design.

2.2.2 BXML
BXML (Binary eXtensible Markup Language) is the first AJAX-based user interface markup languages. It is a proprietary standard by Backbase [7] that runs with all major web browsers and XHTML / DOM-based layout engines. It provides declarative language for cross-browser AJAX development.

2.2.3 XAML
XAML [8] is not just an XML-based user interface markup language, but an application markup language, as the program logic and styles are also embedded in the XAML document. Functionally, it can be seen as a combination of XUL, SVG, CSS, and JavaScript into a single XML schema. Some people are critical of this design, as many standards (such as those already listed) exist for doing these things.

Generally, this approach is based on the fact that most mobile terminals support browsers that are also referred to as micro-browsers. In fact, there is relatively large number of mobile OS and then developers will usually have to redesign an application for different OS when they want to install the application across different mobiles terminal type.

We estimate that an internet connection and a web browser on a mobile terminal aren't sufficient for service provision in pervasive environment which user should directly interact with [9]. A web browser couldn't support all services or their dynamic adaptation according to user profile or other parameters.

2.3 Non-application Oriented Approach

This third approach which corresponds to our concept is similar to the previous one mainly in separation of the UI from the application. In both approaches a user interface description could be used to realize this separation. However, the previous approach uses a web browser as a client to parse the interface description and to display the UI, but this approach allows other specific client to display the UI. Several UI description languages are proposed, among them:

2.3.1 UIML

UIML (User Interface Markup Language) [10] is an XML language for defining user interfaces on computers. Basically UIML tries to reduce the work needed to develop user interfaces. It allows you to describe the user interface in declarative terms (i.e. as text) and abstract it. Abstracting means that you don't exactly specify how the user interface is going to look, but rather what elements are to be shown, and how should they behave. In theory then you could use that description to generate user interfaces for different platforms, like PDAs. In practice, the different capabilities of those different platforms make a complete translation difficult.

2.3.2 UsiXML

UsiXML (which stands for USer Interface eXtensible Markup Language) is a XML-compliant markup language that describes the UI for multiple contexts of use such as Character User Interfaces (CUIs), Graphical User Interfaces, Auditory User Interfaces, and Multimodal User Interfaces.

In other words, interactive applications with different types of interaction techniques, modalities of use, and computing platforms can be described in a way that preserves the design independently from peculiar characteristics of physical computing platform.

The level of abstraction used in UIML and UsiXML is too low that it implies the description of the interface for each target device. In another hand, those languages perform the description of the UI from the graphical display viewpoint but the remote functionalities are not well processed. In the case of UsiXML a local dynamic interaction is proposed but it is not sufficient to make a remote invocation.

In order to deal with a dynamic user interface for pervasive services, independent from application, we are proposing a novel UI description language based on service abstraction model (SAM). In this paper we will focus mainly on SAM.

3 Service Abstraction Model

A service model should facilitate connections between user related parameters, context management features and service provider constraints. In [11], author has proposed an UML model for mobile service. This model is performed for mobile service in order to make easier mobility management (user mobility across different devices) and then service continuity.

In that sense, we propose a Service Abstraction Model (SAM) for assistive pervasive services which contains 3 main parts (fig. 2):

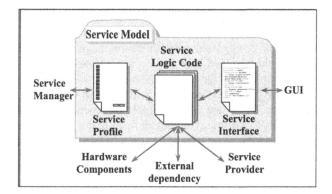

Fig. 2. Service Abstraction Model Components

3.1 Service Logic Code: SLC

SLC is the program code and data that constitutes the dynamic behavior and provides the functions of a service. Obviously, it could have relations with external components depending on the requirement of the service programming itself (hardware components, external dependency, service provider...).

3.2 Service Interface: SI

The SI illustrates the description of the service in term of interactions and functionalities. It answers the following question: how we can use the service. The User-Service Interaction is based on this SI since it represents the interlocutor between user and service core. The SI combines two main interaction types:

- Interaction Interface: represents an abstract description of the GUI of the service. There are several description languages which perform this type of interaction such UIML (as described above).
- Interaction Logic: represents the description of logic functions invoked by the user client during the user-service interaction phase. Its role is equivalent to the WSDL language in web services.

Concretely, SI is a file containing the description of interactions described above. When the service is discovered by the user end-terminal, its SI file is downloaded and interpreted by the suitable software to display the GUI. SI realizes a separation between the core and the interface of the service. We adopted this separation to perform the adaptation and personalization of the user interface to its SP, end-terminal and context of use. Besides, this separation is essential and very useful while creating a dynamic, tangible user interface based on the fusion of different service interface.

3.3 Service Profile: SP

The SP should contain a description of different attributes and proprieties a service can have. Those properties could be classified into two categories:

- Characteristic properties of the service: all what characterizes the service itself. Among them, we mention: service description (what it can do), location, type of contents (video, audio), supported communication mean (i.e. SOAP, RMI…), etc. Other properties could be added here such as security or quality of service aspects (security level, priority regarding emergency cases, Time to Live…).
- Properties describing possible dependency with external entity. Each entity is considered as a category. Those are some example of dependency: 1) Dependency on the user and his profile. Attributes already defined in the UP will be reused here in order to make the link between user and service; i.e. a service could be provided only for specific user or the contrary (e.g. blind person don't need a "lamp" service, or user with very low hand force should be provided with the suitable corresponding service).   2) Dependency on resource. In this part, we can express some service requirement in term of resources the user must have to run the service. Those requirements vary from end-terminal (e.g. minimum screen size) to input/output device (microphone, loudspeaker…). 3) Dependency on other services. This is important to manage service dependency and coexistence; two services requiring both exclusively the use of a resource couldn't coexist actively together. 4) Dependency on external component such as effectors (lamp, elevator) or sensors.

Many other properties could be added, be it alone or as a category. The SP represents the interlocutor between the service platform and the service core especially while filtering and discovery operation.

4 Architecture

We present in this section an over view of our architecture as shown in figure below (Fig. 3). We have presented in this paper one of its component SAM in section 3. The context aware module has the charge of inferring the contextual situation and predicting the potential activity of the user [12]. The user profile module feeds the system with user requirements, preferences and physical capabilities. The interaction handler manage the interaction of the user with the physical interfaces like keypad, trackball, or any other specific device used by the user to interact with his environment. The user environment interaction reasoner is the heart of the system; it combines elements from previous listed modules to make a decision about the look and feel of the GUI and the organisation of the contextual-information-based services. GUI is the outcome of this process and it is adapted to the display device (PDA, tablet PC, TV, etc) and personalized to the user preference and capabilities.

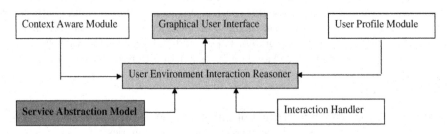

Fig. 3. General architecture

In the section below we will present the development realized and the implementation of our concept within a general demonstrator realized in laboratory conditions. In order to validate this demonstrator and get feedback from end-users, we have exhibited the system during the 5th Workshop on smart homes held in INT in February 2006.

5   General Demonstrator

As illustrated below (Fig. 4), the idea was to build sveral living environments in order to test the service discovery function and its impact onthe dynamicity of the user interface. The designed environment is composed on two indoor living spaces (kitchen and living room), and one outdoor environment representing a bus station service (checking bus timetable on line) and remote action on traffic light crossing button.

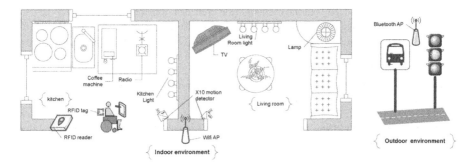

Fig. 4. Experimental Platform

The realized scenario was to fix a tablet PC on the electrical wheelchair of the user and an RFID Tag used to identify the user (user profile). The tablet PC was connected via WIFI network to a home hub and a presence detector was used to locate the user. The mapping between the user preference, his location and the discovered service allowed generating a specific Graphical User Interface in each environment according to each user. In our case three different profiles was experimented.

This demonstrator permitted to highlight the constraint of having a changing GUI for the user. This could not be acceptable in the case of large number of services, maybe due to the difficulties to handle in real condition. Actually we are deploying this platform in a residence of people having severe disabilities (four limbs impairments) in collaboration with local association near INT in order to evaluate the acceptability of a dynamic user interface on different mobile devices (PDA, amrt phone and Tablet PC).

6   Conclusion

This paper described our approach for designing dynamic user interface which could be a solution when considering a pervasive living environement aiming at handling a lerge number of devices and applications. UI is a crucial issue in term of accessibility for independent people where the personnalisation issue is highly requested. We have outlined the difference between existing approaches and have discussed how they could fit with our concept of developing adaptable user interface in assistive living environment. We have detailed our approach which is based on service abstraction methodology and the impact on building dynamic user interface.

We are actually focusing on the implementation of the proposed solution and its integration within a smart home prototype dedicated to dependent people. In short term objective, we are aiming at proposing a markup description language taking into consideration the proposed Service Abstraction Model.

References

1. Hariz, M., Renouard, S., Mokhtari, M.: Designing multimodal interaction for assistive robotic arm, submitted To 10th International Ronference on Rehabilitation Robotics
2. Renouard, S., Menga, D., Brisson, G., Mokhtari, M.: Toward a Document based Model for Multimodal Human-Machine Interaction. In: Proc. ICOST 2005, 3rd International Conference On Smart homes and health Telematics,IOS Press, pp. 204-211, Sherbrooke, CANADA (July 2005) ISBN 1-58603-531-2
3. Fouial, O.: Adaptive service discovery and provision in mobiles environment. Ph.D rapport, ENST Paris, * * Chaari, T., Laforest, F. (April 30, 2004)
4. Chaari, T., Laforest, F.: SEFAGI: Simple Environment For Adaptable Graphical Interfaces -Generating user interfaces for different kinds of terminals. Int. Conf. on Enterprise Information Systems (ICEIS), Paphos - Cyprus (2005)
5. Ariel Pashtan: Mobile Web Services. Cambridge University Press, pp. 38–39 (2005) ISBN 0-521-83049-4
6. XUL web site http://www.mozilla.org/projects/xul/
7. BXML web site http://www.backbase.com
8. XAML web site http://www.xaml.fr/
9. Ghorbel, M., Mokhtari, M., Renouard, S.: A distributed approach for assistive service provision in pervasive environment. In: 4th WMASH '06, pp. 91–100. ACM Press, Los Angeles, USA (2006)
10. UIML web site http://www.uiml.org/
11. Jorstad, I., van Thanh, D., Dustdar, S.: A service continuity layer for mobile services. Wireless Communications and Networking Conference, 2005 IEEE vol. 4, pp. 2300–2305 (March 13-17, 2005)
12. Feki, M.A., Mokhtari, M.: context aware and ontology specification for assistive environment. HWRS-ERS Journal 4(2), P29–32 (2005)

Report on the Impact of a User-Centered Approach and Usability Studies for Designing Mobile and Context-Aware Cognitive Orthosis

Blandine Paccoud, David Pache, Hélène Pigot, and Sylvain Giroux

Laboratoire DOMUS, université de Sherbrooke, Qc, Canada
{blandine.paccoud,david.pache,helene.pigot,
sylvain.giroux}@usherbrooke.ca
http://domus.usherbrooke.ca

Abstract. To foster the mobility of people with cognitive impairements, cognitive orthoses on mobile devices can have a major impact on patients and caregivers quality of life. But patient and caregivers must be implied in the process of design such orthoses. In this paper we describe an iterative process relying on a user-centered approach. On the one hand, it enabled to improve the design and usability of an existing mobile cognitive orthosis. On the other hand, it provide invaluable clues and hints to new context-aware features. Caregivers and patients were deeply involved in the process and two studies were conducted in a real setting with people with schizophrenia and people with head traumas.

Introduction

The ageing population of our modern societies and the future explosion of the number of the persons with cognitive impairments (CI) leads to a real problem in terms of care at home. Pollack [12] in her discourse to the US senate, proposes to use the technology to maintain the autonomy of people and solve a part of this problem. Daily mobility is one crucial part of the autonomy. People with CI encounter specific handicap situations while outside due to planning, memory or attention disorders. Cognitive orthoses aims to alleviate the autonomy of people with CI. Their focus is usuably put on providing simple enough user interfaces.

To foster the mobility of people, many cognitive orthoses have been developed. Neuropage^{TM} uses a pager to send text messages to a pager on a preprogrammed day and hours [4]. Its main advantage is the simplicity of its use. But its features are very limited. Instead of using a pager, Kim proposes the use of PDAs (personal digital assistants)[6]. The use of PDAs allows for richer assistive features.

For instance PEAT, couples a dedicated planner to visual and audiophonic information associated[7]. But its graphical interface is fairly complex, particularly with respect to the use of menus. Thanks to current advances in technology, more sophisticated functionalities will be available without overloading the orthosis usability. For one, Carmien's team mobile application is a context-aware service for memory impaired persons[1]. It offers a context-aware service for memory impaired persons. This application enables the user to obtain information on the next and the previous steps of an

T. Okadome, T. Yamazaki, and M. Mokhtari (Eds.): ICOST 2007, LNCS 4541, pp. 179–187, 2007.

activity. "Opportunity knocks" [9] shows how advanced artificial intelligence systems can provide cognitive assistance to the people with Alzheimer-type dementia.

Since 2003, DOMUS is developing systems on PDAs for people with CI. They supply on the one hand cognitive assistance and tele- monitoring for activities of daily living for people with CI and on the other hand tools to gather ecological medical data [10,2]. A special attention was put on user interfaces [3]. Once we had the first versions of these applications at hand, we started an iterative process based on a user-centered approach and extensive communication with professional caregivers and patients. The aim is to use the full-capacities of current mobile devices to fulfill the ubiquitous needs of user with CI and their caregivers. After a first experimentation with people having schizophrenia, a second iteration is under progress for people having head trauma.

This paper presents the user-centered iterative process and the enhanced and new mobile system features upon those two iterations.

Section §1 presents the current state of the MOBUS application. Section §2 describes how user-centered design enabled to design MOBUS. Section §3 is about the MOBUS configuration system. Last section describes future works identified thanks to the user-centered design.

1   MOBUS, a Mobile Orthosis to Help in Daily-Living Activities and to Keep in Touch with Proxies

Targeted populations. MOBUS targets people with cognitive impairments, such as schizophrenics, people with head traumas (TCC), people who suffered from a cerebral vascular accident (CVA) and persons with Alzheimer's disease. Those four populations share commons effects, but also present some idiosyncrasies (table 1). One challenge in developing a cognitive orthosis for these populations is to take into account differences, and generalize functionalities at the same time to get a single application.

Table 1. comparison table of cognitive impairments encountered among the population with TCC, CVA, or schizophrenia

criterion	DTA	TCC	AVC	schizophrenia
impairements		cognition mood behaviour motor		memory mood behaviour -
age	senior	young adult	adult and senior	young adult
disease evolution	degenerative	stable	stable-enhancement	enhancement under medication

MOBUS. MOBUS is a mobile orthosis involving at least two PDAs, one for the patient and one for each caregiver. It was designed to be as simple as possible. To increase the ubiquity quality, PDA may be replaced by smartphones. A smartphone is a phone with some PDA functionalities. The advantage is the ubiquitous internet connection provided by the GPRS[1] functionality. MOBUS is based on a client-server architecture [3]. MOBUS has four main functionalities.

Activity recalls. This functionality (Fig.1) enables the patient to consult at any time activity recalls. Activity recalls are decided together by the patient and his caregiver. Caregiver enter them using their MOBUS-device. When the patient completes an activity, he validates it on his MOBUS-device. Caregivers can supervise on their PDAs the completion of the activities. They can also manage them: creation, modification, etc.

Symptoms notification. Thanks to the symptom functionality, MOBUS users can record specific symptoms when they feel them, by choosing one in a pre-recorded list and evaluating its intensity on a predefined scale. The application automatically registers date, symptom and intensity. Such is then available for analysis by caregivers and doctors.

Assistance request. If a patient experiments technical or personal problems he can ask for assistance to his caregivers who can accept or decline the request (Fig.1).

Contextual Assistance. The main goal of contextual assistance is to provide information depending on a context designed by the place and the current activity(Fig.1). Information is displayed with pictures or text. When the user goes to a predefined area, MOBUS displays the saved information related to the current activity, such as security rules, orientation help, bus scheduling, etc.

Fig. 1. activity recall, assistance request and contextual-information functionalities of the MOBUS-patient application

2   Use of a User-Centered Design: Development of the MOBUS-System

2.1   User-Centered Design (UCD)

To fulfill the needs of the end-users of MOBUS, we decided to apply a user-centered approach (ISO 13407). This design philosophy places the user as a central pillar of

---

[1] General Packet Radio Services – standard that permits internet connection over a GSM network.

the development process, and allows the final user to be active in the development. Following user-centered design[2], we focus on the particular concerns of the cognitive impaired people. Those concerns are the cognitive factors such as perception, memory and learning. Not only do we need to analyze how users are likely to use our interface, but also to identify how the users behave in their environment.

According to [13] "UCD is [...] considered, in a broad sense, as the practice of the following principles, the active involvement of users for a clear understanding of user and task requirements, iterative design and evaluation, and a multi-disciplinary approach." The goal of UCD is to get a product with a high degree of usability. According to ISO 9241-11 usability is the "extent to which a product can be used by specified users to achieve specified goals with effectiveness, efficiency and satisfaction in a specified context of use." [5].

The user-centered philosophy recommends a process in steps. First one should analyze several criteria such as the final users, their environment, the needed functionalities and their scenarios, the existing similar projects, etc. This analysis leads to the design of the architecture and the thought about metaphors to employ. Design includes a prototype development from a paper prototype to a functional prototype. Finally, the implemented prototype should be evaluated, in the final users' environment to get ecological data. The design and evaluation steps should be iterated long enough to get a final prototype and to implement the final product.[3]

2.2 Goals of the Use of UCD

We highlight four goals of the use of UCD for MOBUS. First, it is to improve MOBUS, by adding some functionalities and perfect the existing ones. Then, it is to satisfy the needs of people with cognitive impairments by collaborating with them and their caregivers. Third, it is to evaluate if MOBUS fulfill these needs. And the last goal is to plan out the future needed functionalities.

2.3 User-Centered Design Adapted to MOBUS

User-centered design as it was previously described needs to be adapted to the population we are targeting [8]. Population with CI may share common deficits, but their individual particularities are also stronger than among a traditional population of users.

The user-centered approach is adapted to the MOBUS application according to three constraints of the targeted population. First, the use of MOBUS needs a couple of users which makes it more difficult to evaluate and adds complexity to the design. Second, the main user has cognitive impairments which often induce lack of introspection. This causes difficulties to evaluate the system and to analyze the needs of the user. Finally, the evaluation is even more complicated as we need ecological data. This implies that the researchers can not be present during the use of MOBUS. This avoids the introduction of biais like using the activity recall because the researcher presence remembers it.

To side step this difficulties, questions which need introspection are asked to the caregivers, and the patients are evaluated in action. Questions may also be asked while

---

[2] http://www.stcsig.org/usability/topics/articles/ucd
[3] http://www.w3.org/WAI/redesign/ucd

the patient and his caregiver are both present, and caregiver helps his patient to answer. Also, to get information about the application use despite the lack of memory and introspection of people with CI, we study logs from the MOBUS system, therefore we collect data such as when the users clicked and where with high precision.

User-centered design advices to use a multi-disciplinary approach. To fulfill this recommendation, we were helped by medical staff (neuropsychologist, psychiatrist, educators) and by the final users themselves. According to the constraints explained above, the multi-disciplinarity of the UCD is well adapted.

2.4   Methodology

UCD advocates three steps iterations (analysis, design and evaluation) until the product satisfies its objectives. We are currently conducting the second iteration of MOBUS. The first iteration enabled us to evaluate the system with a schizophrenic population, whereas the second one is currently done with people who suffered from a vascular accident or a cranial traumatism.

Analysis. A challenge of MOBUS is to be as simple as possible, and adapted to people with cognitive impairments. Analysis of the target populations and their needs takes an important part in this phase. Caregivers were the most valuable persons to whom we were able to ask our questions and with whom we were able to discuss about the functionality needed by people with CI. So we organized appointments with them, for each of the experimentations. Finding useful functionalities was possible via scenarios. We told the caregivers what kind of functionalities we could develop, and by thinking to their patients, they could give us scenarios around this functionalities and how we could improve them. They may then propose new functionalities and we were able to tell them if it was technically realizable or not. We did not recruit participants ourselves. The caregivers were in charge of recruitment because first they know better their patients and their abilities to participate to the research and second it is hard to recruit people with CI as the population is limited in number.

Design. During the design phase, we thought to the functionalities studied with the caregivers, taking into account their advices. We worked at designing those functionalities, still with the constraints to have a very simple interface on a small screen. We also met the caregivers during this phase, in order to validate modifications. Then we could implement a simple prototype, to be evaluated with the final-users.

Evaluation. To evaluate technology with people with cognitive impairments is difficult and restrictive (§2.3). We had to adapt the evaluations to each of the targeted populations. We organized two different experimentations, one for each cycle. The first experimentation with three schizophrenics lasted a week. A first individual appointment was organized with the participants, who already knew about the experimentation. Researchers were able to explain the evaluation in details, talk with the participants, fill in questionnaires and get used to the equipment. During a week, participants used the mobile orthosis and tested the functionalities. At the end, they gave their feedbacks. A week was enough because the targeted users were young and used to technology. They quickly learned how to use MOBUS.

The second experimentation is under process. MOBUS is evaluated with people who suffered from a vascular accident and people with cranial traumatism. We evaluate MOBUS with four of them presently. We decided to add a learning phase, where we teach the users how to use MOBUS. There are two reasons to this addition. First, participants are not confident with this kind of technology because they do not use it daily. Second, they have problems to include the orthosis in their habits and to remember to use it, because they need time to get used to novelty. By using scenarios, the patient can learn how to handle the system. On our side, we can take notes and control the evolution of the learning, appointments after appointments. We planed the learning of the system on four to six meetings of twenty to thirty minutes each. When a participant could remember enough information, then he may be able to use the system for a week, daily.

Data collection is a major step in the evaluation phase. We elaborated several questionnaires, forms and interviews to be applied all along the experimentation. Questionnaires are filled before the experimentation, to get to know the participants habits, after the learning phase (for the second experimentation) to get their feelings about it, and finally after the experimentation, to evaluate the interface of the system, and collect their ideas and critics about the evaluation and MOBUS. Forms are mainly used during the learning phase. During this phase, one of the researchers exposes a scenario to a patient, and the patient has to use MOBUS as if he were in that situation (i.e.: You are at home and you want to know what your next activity is, what do you do?). The other researcher notes the observation and critics of the user on special forms. This enables to study how MOBUS is used and how the user retains information all along the learning phase. Finally, we have two interviews. The first one is scheduled before the experimentation to answer the user's questions and reassure him. The final one takes place after the whole evaluation. It is the major interview because contrarily to questionnaires with static and bounded questions, here, there is just a guiding line which leaves the patients free to give their opinion. It is easier for them to expose their critics and ideas.

2.5   Results

Thanks to the UCD, we fulfill our first three goals : develop and improve MOBUS functionalities, with the collaboration of caregivers and people with CI. The collaboration with a psychiatrist led us to the implementation of symptoms evaluation functionality. Patients often have side-effects due to medication or to the cause of their impairment (cranial traumatism, schizophrenia...). But because it is hard to trace those effects, it was difficult to associate them to a particular moment of the day or to a specific medication. We implemented the symptoms evaluation, allowing to evaluate a symptom by its intensity, and to trace those symptoms.

To simplify the understanding of activity recalls, it was advised to use colors. A green recall is an activity in time, a yellow one means the activity needs to be started, and a red one is a late activity. At the beginning, the change of color was automatic (at predefined percentages of the activity). But this was too restrictive. We modified the functionality in order to enable the caregivers to configure when the recall needs to change its colors.

The collaboration with the caregivers enabled us to refine the contextual assistance. After presenting hypothetic scenarios using GPS and the limits of the device, professional caregivers gave us three kinds of scenarios more valuable: recommendation scenario, instruction scenario, and specific information scenario. The recommendation scenario is based on a simple observation: in their daily activities, people with CI sometimes need recommendations in order to increase their security. For example: someone often goes to the stadium with his child to watch a football game. Caregivers want to remind some recommendations in this specific context such as waiting for everybody to leave the arena.

The instruction scenario provides information on the accomplishment of a specific daily activity. For example, when the person often patronizes a building like a rehabilitation center, and can't memorize how to access a specific office, the application will provide a map to help him find the way. This notion is akin to the research of Carmien [1] which suggests the definition of daily scenarios to the users and gives step-by-step information to realise them.

Sometimes people need specific information like bus timetable, or direction in some area where choices are difficult like crossroads. The information can not be associated to a particular time-defined activity. The location is precise enough to define the information to be displayed.

3   Configuration

Adaptation to targeted population is one of the most important goals of the research team. To be able to fulfill the needs of any user, we chose to enable high configuration of the system. MOBUS can be configured according to two types of configuration. The first one concerns the individual needs of the users whereas the second concerns the general needs of a population.

Individual configuration enables MOBUS to be adapted to the user and its capacities: for example, some users will be confident enough with only two activity recalls as shown on Figure 1 whereas others would manage with four or five recalls on the screen. MOBUS can also be adapted on several aspects: colors of recalls of activities, number of recalls, type of symptoms, area of contextual information, etc.

On the other hand, MOBUS can be configured to show only some of its functionalities. As it was introduced about the targeted populations, users may have common or different consequences of their disabilities. It is very difficult to draw up a typical portrait of the targeted population[11]. So functionalities may be available for some of them but not for the others. A user with schizophrenia may not need orientation help, while someone with a cranial traumatism may need it.

4   Future Works

To achieve the third goal, we need to study the results of the experimentation. But some technical improvement are already identified. For example, the development of the symptom history interface for the caregivers has already started and will be tested in the third process iteration. The next step of the development includes several points of research. First in order to increase the efficiency of the activities recall, the application needs an intelligent planning. As for the contextual information assistance, the contextual knowledge will be improved by using web services including weather information and transportation one.

Finally, a configuration system will have to be developed on standard PC platforms. The limitation of the smartphone or PDA (interface and weaknesses of computation resource) and the sophistication of the patient application lead to a difficult gathering of the functionality design and to an increasingly complex interface for the helper system. This development enables to establish basis of future work on statistical exploitation of the data collected by MOBUS.

5   Conclusion

We have developed an application in order to assist persons with cognitive impairments in their daily activities. The design and features of the mobile application is results from a user-centered approach. Two usability studies were performed[4]. Using the UCD approach enables us to fulfill more adequately the needs to the users and to ensure that people with cognitive impairments are able to use MOBUS.

References

1. Carmien, S.: End user programming and context responsiveness in handheld prompting systems for persons with cognitive disabilities and caregivers. In: Conference on Human Factors in Computing Systems CHI'05, Extended Abstracts on Human Factors in Computing Systems, pp. 1252–1255, présentation de MAPS (April 2-7, 2005)
2. Giroux, S., Pigot, H., Mayers, A.: Indoors pervasive computing and outdoors mobile computing for assisted cognition and telemonitoring. In: Computers Helping People with Special Needs, vol. 31, pp. 953–960. Springer, Berlin, Heidelberg (2004)
3. Giroux, S., Pigot, H., Moreau, J.-F., Savary, J.-P.: Distributed mobile services and interfaces for people suffering from cognitive deficits. pp. 544–554. Ismail Khalil Ibrahim (2006)
4. Hersh, N., Treadgold, L.: Neuropage: the rehabilitation of memory dysfunction by prosthetic memory and cueing. Neurorehabilitation 4, 465–486 (1994)
5. Jokela, T., Iivari, N., Matero, J., Karukka, M.: The standard of user-centered design and the standard definition of usability: analyzing ISO 13407 against ISO 9241-11. In: CLIHC '03: Proceedings of the Latin American conference on Human-computer interaction, pp. 53–60. ACM Press, New York, NY, USA (2003)
6. Kim, H.J., Burke, D.T., Dowds, M.M., Georges, J.: Utility of a microcomputer as an external memory aid for a memory-impaired head injury patient during in-patient rehabilitation. Brain Injury, vol. 13(2), pp. 147–150, Premiers test cliniques avec un PDA comme orthése cognitive (1999)
7. Levinson, R.: A custon-fitting cognitive orthotic that provides automatic planning and cueing assistance. In: Technology and Persons with Disabilitites Conference, Présentation littéraire des aspect planification de PEAT (2004)
8. McGrenere, J., Sullivan, J., Baecker, R.M.: Designing technology for people with cognitive impairments. In: CHI '06: CHI '06 extended abstracts on Human factors in computing systems, pp. 1635–1638. ACM Press, New York, NY, USA (2006)

---

[4] The second one will be completed by the end of February 2007.

9. Patterson, D.J., Liao, L., Gajos, K., Collier, M., Livic, N., Olson, K., Wang, S., Fox, D., Kautz, H.: Opportunity knocks: A system to provide cognitive assistance with transportation services. In: UbiComp 2004, LNCS, vol. 3205, pp. 433–450, Davies, N (2004)
10. Pigot, H., Giroux, S.: Keeping in touch with cognitively impaired people: How mobile devices can improve medical and cognitive supervision. IOS Press, Amsterdam (September 15–17, 2004)
11. Pigot, H., Savary, J.P., Metzger, J.L., Rochon, A., Beaulieu, M.: Advanced technology guidelines to fulfill the needs of the cognitively impaired population. In: Giroux, S., Pigot, H. (eds.) 3rd International Conference On Smart Homes and health Telematic (ICOST), July 4-6, 2005. Assistive Technology Research Series, pp. 25–32. IOS Press, Amsterdam (2005)
12. Pollack, M.E.: Assistive technology for aging populations. In: Special committee on aging united states senate (2004) note: `http://www.eecs.umich.edu/pollackm/Pollack-web.files/senate-testimony.pdf`
13. Vredenburg, K., Mao, J., Smith, P.W., Carey, T.: A survey of user-centered design practice. In: CHI '02: Proceedings of the SIGCHI conference on Human factors in computing systems, pp. 471–478. ACM Press, New York, NY, USA (2002)

Context Aware Life Pattern Prediction Using Fuzzy-State Q-Learning

Mohamed Ali Feki[1], Sang Wan Lee[2], Zeungnam Bien[2], and Mounir Mokhtari[1]

[1] Institut Nationale des Télécommunications, 9 rue Charles Fourier 91011 Evry, France
[2] Korea Advanced Institute of Science and Technology, Daejeon 305-701, Republic of Korea
{mohamedali.fki,mounir.mokhtari}@int-evry.fr,
zbien@ee.kaist.ac.kr, bigbean@ctrsys.kaist.ac.kr

Abstract. In an Assistive Eenvironment (AE), explicit/obtrusive interfaces for human/computer interaction can demand exclusive user attention and, often, replacement of them with implicit ones embedded into real-world artifacts for intuitive and unobtrusive use is desirable. As a part of solution, Context Aware can be utilized to recognize current context situation from a combination of low-level sensed contexts. Assuming the current context recognized, this paper tackles the next logical step of "the prediction of future contexts". This information allows the system to know patterns and their interrelations in user behaviour, which are not apparent at the lower levels of raw sensor data. The present paper analyzes prerequisites for user-centred prediction of future context and presents an algorithm for autonomous context recognition and prediction, based on our proposed Fuzzy-State Q- Learning technique as well as on some established methods for data-based prediction.

Keywords: Context aware, assistive environment, prediction, Fuzzy-State Q-learning.

1 Introduction

Assistive environment (AE) is a physical space that can gather one or many people with disabilities and their needed Assistive technology (AT), and should be able to provide users with accessible services and activities they want to perform using existing (AT) and emerging technologies. Assistive environment is rapidly evolving from a proven concept to a practical reality. In deed, advances in smart devices, mobile wireless communications, sensor networks, pervasive computing, machine learning, middleware and agent technologies, and human computer interfaces have made the dream of assistive environment (AE) a reality. An important characteristic of such an intelligent, ubiquitous computing and communication paradigm lies in the autonomous and pro-active interaction of smart devices used for tracking user's' important contexts such as current and near-future locations as well as activities. *Context awareness* is indeed a key to build a smart environment and associated applications. As for example, the embedded pressure sensors in the Aware Home [1] capture inhabitants' footfalls, and the system (i.e., smart home) uses these data for position

T. Okadome, T. Yamazaki, and M. Mokhtari (Eds.): ICOST 2007, LNCS 4541, pp. 188–195, 2007.

tracking and pedestrian recognition. The Neural Network House [2], the Intelligent Home [3], the Intelligent House_n [4] and the MavHome1 [5] projects focus on the development of adaptive control of home environments by also anticipating the location, routes and activities of the inhabitants. Intelligent prediction of such contexts helps in efficient triggering of mobility-aware services. From this background we have lead the interest of applying similar predictive technology in order to trigger assistive –aware services and ensure their adaptation to people with disabilities.

In another hand, in assistive environment, Context prediction ultimate goal considers making possible proactive devices and device interfaces that go some way towards the provision of the user's intentional service with an optimal delay, or to anticipate some assisted user's tasks in order to automatically perform it. Context Prediction seems more interesting area when applied to assistive environment especially if it could bring right environment organisation in some critical situation (dangerous). Some research perspectives on possible context architectures provide helpful insights. In particular, [6] reasons that context monitoring and reasoning are resource consuming, and that to address this issue perhaps distributed and peer-to-peer approaches could be used. Authors also raise the important issues of prediction sharing, where individuals or their environments may have access to the contexts, context histories and/or context predictions of others. In their research, Petzold et al. focus also on context prediction based on previous behaviour patterns. Their proposed prediction algorithms originate in branch prediction techniques (known from the area of processor architecture), which are transformed to handle context prediction [7]. Their evaluation shows that the proposed context predictors exhibit issues in their ability to learn complex patterns.

Applied to assistive environment, The Aware-System [8] seeks to reduce the time difference between current controlled communication devices and actual conversation by utilizing the conversational prediction model developed by Norman Alm [9]. Unfortunately, this system addresses an only conversational issue which is one single parameter of communication services and in the way it is designed it can not be extensible to predict many users' services such as home control, transportation, emergency, etc.

To deal with context aware application, several frameworks have addressed the problem of context redesign and modelling. [10, 11, 12]. However, none of them has focused the stress on context prediction. In SOCAM [13] and Cobra [14], which are based web semantic modelling for inferring high-level context from raw information stored in a data repository, authors have applied a web semantic ontology for designing context aware appliances for smart environment. Their approach has many advantages in terms of knowledge sharing and reasoning with contextual situation, but does nor support assistive services that need to be adapted into handheld and resource-limited end users devices. A recent work elaborated by [15], uses Markov model patterns to allows prediction of abstract contexts in order to help computer systems to proactively prepare for future situations. Their theoretical model is strongly described but does not offer simple implementation API in order to be applied to assistive environment.

The contribution of this paper deals with a new algorithm for context aware assistive services prediction and its integration into a context aware framework supporting acquisition, interpretation, storage and reasoning about relevant context. The outcome of this framework is a set of contextual situation that will be considered as input for our reference context predictive algorithm in order to predict with the right accuracy future contexts recognized with uncertain data.

2   Methodology

2.1   Markov Decision Process and Uncertainty

Our problem is to discover human's life pattern based on the sequential tasks. Thus, Markov process would be an appropriate solution, since it can be described by the states and their transition and assumed that the transition from the current state to the next state depends upon only current one. Moreover, it enable us to decide which state is the most likely one next time, i.e., gives us an optimal strategy for prediction. Thus, it is an appropriate way to optimize the behavior with Dynamic Programming or Q-learning [16], [17].

For this problem, however, we cannot apply Dynamic Programming or Q-learning technique directly because the states often lie in uncertain situation and in that case such methodologies may not apply. It goes without saying that probabilistic approach for Markov Decision Process, such as HMM, has an inherent danger in handling uncertain situation, due to the possible conceptual conflict between uncertainty and probability[18], [19]. Thus, it is strongly required to bridge the gap between the existence of the state and the uncertainty in their nature. The key lies in fuzzy theory, because fuzzy concept is a strong mathematical tool for dealing with uncertainty. There were several studies [16], [20], [21] on how to handle the situation that the states lie in 'uncertain' situation in the field of fuzzy-related Q-learning. To deal with this problem, Modified Q-Learning Method with Fuzzy State Division [16] claims an appropriate solution based on the expert's knowledge, which seems apt for our problem. However, we have no expert to provide the life pattern, i.e., we have no idea on how to categorize it.

Thus, in the next section, in consideration of this fuzzy and categorizing techniques, we propose a combined approach, Fuzzy-State Q-Learning, to conduct the underlying uncertainty to the state based on the categorization by fuzzy-clustering technique.

2.2   Fuzzy-State Q-Learning

Fuzzy-State Q-Learning is an extended version of Modified Q-Learning Method with Fuzzy State Division [16], aiming not only to deal with states under uncertain situation, but also to exploit concurrent pattern from them.

The procedure starts with the initialization, in which the two parameters for Q-learning and the number of fuzzy partition C are to be set. The number of fuzzy partition decides how many linguistic descriptions are needed to reflect model's uncertainty. For example, if there is two kinds of attributes, temperature and time, then in *Step2*, by setting $C=3$ we will obtain three kinds of fuzzy partitions from the given data in the space of two attributes. *Step3* mostly follows a general process of Modified Q-Learning Method with Fuzzy State Division; Starting from the current state, selecting an action by moving to the next state for maximizing an expected reward in fuzzy manner, and then update the current state's Q-value.

Procedure : Fuzzy-State Q-Learning

Step1. Initialization

  Decide the parameters, learning rate α , discount factor γ , and number of fuzzy partition C.

Step2. Fuzzy Partition for the state with uncertainty

  For given attributes with uncertainty $X = \{x_1, \cdots, x_N\}$,

  construct C fuzzy partitions v_i $(i=1, \cdots, C)$
    by any fuzzy clustering method with a normalized membership, $\mu_i(x_k) = \dfrac{1/\|x_k - v_i\|^2}{\sum\limits_{j=1}^{c} 1/\|x_k - v_j\|^2}$.

Step3. Fuzzy Q-Learning

  Initialize Q-values, $Q_j(s,a)$ for $\forall s, a, j$.

  Do

    $S \leftarrow$ current state

    Select the action $a^* = \arg\max\limits_{a} FQ(s,a)$,where $FQ(s,a) = \sum\limits_{j=1}^{c} \mu_j(x) Q_j(s,a)$

    Execute the action by moving to the new state s', and get reward r.

    Update $Q_j(s,a) \leftarrow Q_j(s,a) + \alpha_j \left\{ r + \gamma \max\limits_{a'} FQ(s',a') - Q_j(s,a) \right\}$ for $\forall j$
      ,where $\alpha_j = \alpha \cdot \mu_j(x)$.

  until Q-values converge or maximum number of trials are reached.

Fig. 1. Fuzzy-State Q-Learning : Modified Fuzzy Q-Learning Algorithm

We use the term "fuzzy Q-value", instead of Q-value. Since state lies in an uncertain situation categorized in *Step2*, each state's fuzzy Q-value is consisted by C kinds of sub-Q-values, each of which corresponds to the fuzzy membership value. Thus, the fuzzy Q-value for the action can be calculated by the linear combination of them. And then the update is conducted by the general rule of Modified Q-Learning Method with Fuzzy State Division, except the fact that the learning parameter is controlled by the corresponding fuzzy membership value, i.e., certainty.

Example: Fig.2 helps an intuitive understanding. Suppose that our task is to build a service robot caring "George". If robots' observing states could be defined by "drinking coffee", "watching TV", and "having a meal", then the corresponding action would be interpreted as "serving coffee", "turning on TV", and "serving a meal". Again suppose that each state lies in time and temperature lines, then our George's life pattern can be described in two dimensional space defined by them. Thus, as a result of Step 2, we will get the categorization of the George's life pattern, such as "When cold morning, George usually drinks coffee.", "When cold afternoon, George likes to watch television.", "When hot morning, George likes to watch television.", and "When hot evening, George usually enjoy a meal.". In Step3, from the habits obtained from George, we are able to make a connection among them. Indeed, it will enable the robot to predict what George wants next, based on the previous action he took. As a consequence, serving coffee on 15:00 O'clock under the temperature of 5 can be quantified as the linear combination of "serving coffee on cold morning with certainty level of 0.2", "serving coffee on cold afternoon with certainty level of 0.7", "serving coffee on hot morning with certainty level of 0.02", and "serving coffee on hot afternoon with certainty level of 0.08".

Example

State		Action		Context attributes
				with uncertainty
S_{00001} : *meal served*	(o)	a_1 : *serving a meal*		
S_{00101} : *meal served & coffee served*	(△)	a_2 : *turning on TV*		*time and temperature*
S_{01101} : *meal served, coffee served*	(+)	a_3 : *serving coffee*		
⋮ *& PC turned on*		⋮		

Fuzzy partition for the state   *Linguistic description for each membership function*

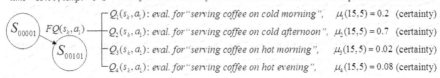

┌ μ_1 : *"cold morning"*

├ μ_2 : *"cold afternoon"*

├ μ_3 : *"hot morning"*

└ μ_4 : *"hot evening"*

Linguistic description for fuzzy-state transition

"*time* = 15:00, *temp.* = 5°C". *Breakfast were already served. Which service is the most probable one next?*

$Q_1(s_k, a_i)$: *eval. for* "*serving coffee on cold morning*", $\mu_1(15,5) = 0.2$ (certainty)

$Q_2(s_k, a_i)$: *eval. for* "*serving coffee on cold afternoon*", $\mu_2(15,5) = 0.7$ (certainty)

$Q_3(s_k, a_i)$: *eval. for* "*serving coffee on hot morning*", $\mu_3(15,5) = 0.02$ (certainty)

$Q_4(s_k, a_i)$: *eval. for* "*serving coffee on hot evening*", $\mu_4(15,5) = 0.08$ (certainty)

S_{00001} $FQ(s_k, a_1)$ S_{00101}

Fig. 2. Elements of Fuzzy-state Q-Learning and their linguistic descriptions

2.3 Remark on Convergence

If we substitute fuzzy Q-value by Q-value in Fig.1, then the learning equation will be the same form as the one in Q-learning. Because we use the normalized fuzzy memberships, whose summation is one, the linear combination of sub-Q-values with fuzzy membership coefficients becomes convex hull. Therefore, the fuzzy Q-value lies in the convex hull consisted by sub-Q-values

$$FQ(s,a) \in convex\ hull\ of\ Q_j(s,a) \cdot \tag{1}$$

It means that fuzzy Q-value is equal of less than the maximum value of sub-Q-values, such that

$$|FQ(s,a)| \leq \max_j |Q_j(s,a)| \cdot \tag{2}$$

By substituting with the upper bound of (2), we get

$$\Delta Q_j(s,a) \le \alpha\mu_j \left\{ r + \gamma \max_{\alpha'} \max_j Q_j(s',a') \right\}$$

$$\le \alpha \left\{ r + \gamma \max_{\alpha',j} Q_j(s',a') \right\}. \tag{3}$$

If we consider j-th sub-Q-values as general term of Q-value, then

$$\Delta Q(s,a) \le \alpha \left\{ r + \gamma \max_{\alpha'} Q(s',a') \right\}. \tag{4}$$

Now we are back to general Q-learning problem in which the convergence is proved [17]. Thus, fuzzy Q-value in the learning rule of Fig.1 is a convergent sequence.

3   Prototype

The following scenario illustrated in the figure below is aiming at validating our concept into an integrated demonstrator settled in the lab environment. The application consists on providing a user, having an electrical wheelchair; three different unknown contexts specified with :(outdoor-bus station, indoor-home_room1 and indoor-home_room2). The mobile HMI, based on a PDA, a Tablet PC and a smart phone mounted on the wheelchair, discovers the services according to corresponding context and to the user profile formalised in an RFID tag. Three types of services are provided: bus station schedule and traffic light status in context 1, lights in context 2 and lights and TV in context 3. In addition the HMI permanently controls the arm robot which is mounted on the wheelchair. An extension to wheelchair control is also planned. The aim of our predictive algorithm is to interact with a context framework developed with JAVA OSGI bundle and predict the next context (so the next task or service) based on the current context situation provided by the context framework.

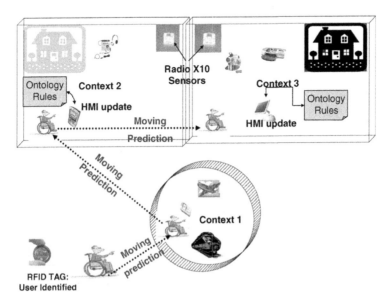

Fig. 3. Prototype scenario

4 Simulation Results

We consider 5 kinds of task ; "Serving a meal", "Turn on TV", "Serving Coffee/tea", "Turn on PC", "Preparing a bath". Since each task can be executed several times, we define binary state description, where each of the state is represented by 5-bit number. Thus we have 32 states.

Lifepattern DB, which consists of 157 tasks, was collected during 2 weeks. 78 tasks were used for training, while 79 used for testing. The results are shown in Table1. It is noticeable that there are unavoidable cases causing errors, e.g., several tasks are executed in the same time unit, or different tasks were executed on same time but different day. Despite of these difficulties, fuzzy-state Q-learning shows 72.1% success rate, whereas Q-learning shows the performance of 49.3%.

Table 1. Simulation results: Life pattern prediction

Algorithm	Parameter setting	Success Rate (%)
Q-learning	$\alpha = 0.1$, $\gamma = 0.07$, $r = \pm 1$	49.3
Fuzzy-state Q-Learning	$\alpha = 0.1$, $\gamma = 0.07$, $r = \pm 1$, $C = 30$	**72.1**

5 Conclusion

This paper seeks to figure out methodology design in order to deal with context prediction applied to assistive environment (AE). We outlined the transformation of the context prediction challenge into different mathematic solution and argued the convergence of fuzzy Q-value.

We presented also an architecture that supports the context prediction and its integration with assistive environment domain.

We are actually dealing with implementation and integration of the concept into real living environment (residence of people with disability).

References

1. Orr, R.J., Abowd, G.D.: The Smart Floor: A Mechanism for Natural User Identification and Tracking. In: Proceedings of 2000 Conference on Human Factors in Computing Systems (CHI 2000), ACM Press, NY (2000)
2. Mozer, M.C.: The Neural Network House: An Environment that Adapts to its Inhavitants. In: Proc. of the American Association for Artificial Intelligence Spring Symposium on Intelligent Environments, pp. 110–114 (1998)
3. Lesser, et al.: The Intelligent Home Testbed. In: Proc. of Autonomy Control Software Workshop (January 1999)
4. House_n Living Laboratory Introduction
   http://architecture.mit.edu/house_n/web/publications
5. Das, S.K., Cook, D.J., Bhattacharya, A., Hierman, E., Lin, T.Y.: The Role of Prediction Algorithms in the MAVHome Smart Home Architecture. IEEE Wireless Communications, Special Issue Smart Homes 9(6), 77–84 (2002)

6. Nurmi, P., Martin, M., Flanagan, J.A.: Enabling Proactiveness through Context Prediction. In: Proceedings of the Workshop on Context Awareness for Proactive Systems, Helsinki (2005)

7. Petzold, J., Bagci, F., et al.: Global and Local State Context Prediction. Artificial Intelligence in Mobile Systems 2003 (AIMS 2003) in Conjunction with the Fifth International Conference on Ubiquitous Computing 2003, Seattle, USA (2003)

8. Adams, L., Hunt, L., Moore, M.: The "Aware- System" - Prototyping an Augmentative Communication Interface, presented at RESNA 2003 (2003)

9. Alm, N., Arnott, J.L., Newell, A.F.: Prediction and Conversational Momentum in an Augmentative Communication System. Communications of the ACM 35, 46–57 (1992)

10. Dey, A.K.: Understanding and Using Context. Personal and Ubiquitous Computing, Special issue on Situated Interaction and Ubiquitous Computing (2001)

11. J. E. Bardram, UbiHealth 2003: The 2nd International Workshop on Ubiquitous Computing for Pervasive Healthcare Applications, Seattle, Washington, October 12, part of the UbiComp 2003 Conference http://www.healthcare.pervasive.dk/ubicomp2003/papers/

12. Schmidt, A.: Ubiquitous Computing - Computing in Context. Ph.D. Dissertation, Department of Computer Science, Lancaster University (November 2002)

13. Gu, T., Wang, X.H., Pung, H.K., Zhang, D.Q.: An ontology-based context model in intelligent environments. In: Proceedings of the Communication Networks and Distributed Systems Modeling and Simulation Conference (CNDS'04) (January 2004)

14. Chen,H., et al.: Intelligent Agents Meet the Semantic Web in Smart Spaces. Article, IEEE Internet Computing (November 2004)

15. Mayrhofer, R.: An Architecture for Context Prediction. PhD thesis, Johannes Kepler University of Linz, Austria (October 2004)

16. Yoichiro, M.: Modified Q-Learning Method with Fuzzy State Division and Adaptive Rewards. In: Proc. of IEEE International Conference on Fuzzy Systems, pp. 1556–1561(2002)

17. Christopher, J.C.H.W., Peter, D.: Q-Learning. Machine Learning 8, 279–292 (1992)

18. Bezdek, J.C.: Fuzziness vs. Probability - Again. IEEE Trans. Fuzzy Systems 2(1), 1–3 (1994)

19. B. Kosko.: The probability Mopnopoly. IEEE Transactions on Fuzzy Systems. vol. 2(1) (1994)

20. Hamid, R.B.: Fuzzy Q-learning: A New Approach for Fuzzy Dynamic Programming. IEEE World Congress on Computational Intelligence, pp. 486–491 (1994)

21. Suh, I.H., Kim, J.-H., Frank Rhee, C.-H.: Fuzzy Q-learning for Autonomous Robot Systems. In: Proc. of IEEE International Conference on Neural Network, pp. 1738–1743 (1997)

Distributed Vision-Based Accident Management for Assisted Living

Hamid Aghajan[1], Juan Carlos Augusto[2], Chen Wu[1], Paul McCullagh[2], and Julie-Ann Walkden[3]

[1] Wireless Sensor Networks Lab, Stanford University, USA
[2] School of Computing and Mathematics, University of Ulster, UK
[3] Ulster Community Hospitals Trust, UK

Abstract. We consider the problem of assisting vulnerable people and their carers to reduce the occurrence, and concomitant consequences, of accidents in the home. A wireless sensor network employing multiple sensing and event detection modalities and distributed processing is proposed for smart home monitoring applications. Distributed vision-based analysis is used to detect occupant's posture, and features from multiple cameras are merged through a collaborative reasoning function to determine significant events. The ambient assistance provided will assume minimal expectations on the technology people have to directly interact with. Vision-based technology is coupled with AI-based algorithms in such a way that occupants do not have to wear sensors, other than an unobtrusive identification badge, or learn and remember to use a specific device. In addition the system can assess situations, anticipate problems, produce alerts, advise carers and provide explanations.

1 Introduction

A growing application in which sensor networks can have a significant impact on the quality of the service is monitored care and assisted living. Traditional solutions in the 'Smart Home' domain limit the independence of the person receiving care. However, networked sensors can collaborate to process and make deductions from the acquired data and provide the user with access to continuous or selective observations of the environment. Thus these networks have the potential to dramatically increase our ability to develop user-centric applications to monitor and prevent harmful events.

Whilst there had been research conducted in the use of Smart Homes to enhance the quality of life for senile dementia patients [1], users have to explicitly interact or wear different devices. Here we report a framework which overcomes those strong assumptions but still can help with care provision by detecting hazardous events and notifying carers. We focus on the propensity of the *elderly to suffer falls* [2]: a) they are a major cause of disability and loss of independence, causing fractures, head injuries, pressure sores, depression and social isolation, b) they are the main cause of accidental death in this group. Approximately 30 percent of the people over 65 years living in the community face an accidental

T. Okadome, T. Yamazaki, and M. Mokhtari (Eds.): ICOST 2007, LNCS 4541, pp. 196–205, 2007.

fall each year, while the number is higher in institutions. Although less than 10 percent results in a fracture, a fifth of incidents require medical attention [3], posing a significant financial burden to the health service.

The wireless sensor network developed in this work employs visual information acquired by a set of image sensors, as well as a user badge that provides accelerometer signaling, position sensing capabilities and voice communication with the care center as the basis for making interpretations about the status of persons under care. For example, in a monitoring application, the location, duration of stay at one position, sudden moves, and body posture of the occupant can be used in combination with rule-based logic to evaluate various hypotheses about their well being. Such multi-modal sensing and network-based design can also be extended by other sensory devices making various user-centric measurements. The frequency and sequence of the measurements taken by the different elements in the network can be determined adaptively according to the nature of the expected events that occur around the user. Images can be analyzed to assess whether the person is moving appropriately and performing activities of daily living or whether a fall or undue inactivity has occurred.

2   System Model

The system's configuration is inspired by the Smart Home care network developed for elderly monitoring [4], [5]. Multi-camera vision processing is employed to analyze the occupant's posture before, during, and after an alert, which may be generated by a user badge equipped with accelerometers. This analysis results in an assessment of the situation, and directs the system to produce the proper report for the call center. Our prototype design consists of three to five cameras installed on the wall. A voice transmission circuit allows the system to create a voice link between the user and the care center automatically or by user's demand. Through distributed scene analysis, each camera node processes the scene independently. Collaborative reasoning among the cameras results in a final decision about the state of the user (Figure 1(a)). Since each single camera cannot be expected to always extract all the desired information from the scene, utilizing a multi-camera data fusion technique will provide complementary information when needed, resulting in a more accurate understanding of the scene. As a result, the system will be able to make a more reliable decision, create a more efficient report, and reduce the number of false alarms sent to the central monitoring center (Figure 1(b)).

3   Vision-Based Posture Analysis

Employment of multiple image sensors allows for covering different areas of the environment as well as having different viewing angles to the same event. Due to the self-occlusive nature of human body and occlusions from the indoor environment, observation from a single camera is prone to ambiguities. Therefore, collaboration between the network nodes can be used to combine attributes

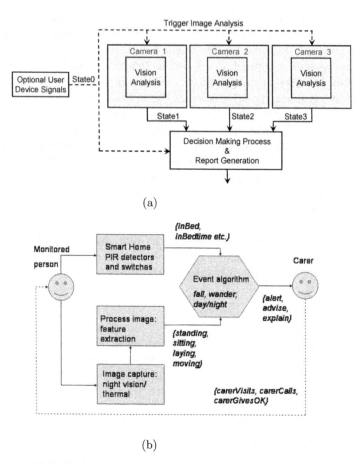

Fig. 1. (a) Decision fusion architecture. (b) Block diagram of the system.

and derive more confident deductions about the events and status of the user. In-node image processing enabled by today's embedded processors helps mitigate the need to transfer raw images across the network, and hence reduces the constraints on bandwidth management of wireless networks. It also allows for addressing user privacy concerns by avoiding unnecessary image transmission to the care center. The vision processing module also creates silhouette or model-fitted representations of the user for transmission to the call center to further alleviate privacy concerns (see Fig. 2(b),(c)).

Having multiple cameras in the solution also enables operation even if the user forgets to carry the badge with accelerometers. In such a case the user-based device for detecting a fall is not available. However, with the help of the cameras, which could be activated frequently or upon a timeout if they do not hear from the user badge, useful information about the situation of the user can be extracted. The status or posture of the user is analyzed through the layered architecture illustrated in Fig. 2(a). Some example images from falls and other indoor activities are shown in Fig. 2(b). Image features of the foreground are extracted for further analysis of human postures. Those features may include motion vectors, color distribution, body part segmentation, etc (Fig. 2(c)). Posture elements are derived from image features, which are more distinctive as descriptions for postures, such as global motion, motion and color of segments [6], [7]. Collaboration between cameras is actively pursued in decision making from features to posture elements, and from posture elements to posture descriptions.

4   AI-Assisted Decision Making

Smart Homes technology greatly relies on context-awareness reasoning to relate contexts with accurate prediction and sensible recommendations. Once an image has been identified as interesting by the detection system, it can be combined with previous images and other contextual information (location and time constraints) to achieve a higher level view of the developing scenario.

Our situation diagnostic system uses a notation [8] that allows a simple specification with emphasis on the causal relationships between different elements of a Smart Home scenario. The system distinguishes between *independent* (S_I) and *dependent* (S_D) states. The former are states whose value can change at any time, outside the control of the system and therefore it cannot be anticipated. They will represent the state of sensors being stimulated by activities in the house. Dependent states can be expressed in the heads of causal rules as an effect of the system reaching other states. There are two kinds of *causal rules*:

$$S_1 \sqcap S_2 \sqcap \cdots \sqcap S_n \rightsquigarrow S \ (Same\text{-}time \ rules)$$

$$S_1 \sqcap S_2 \sqcap \cdots \sqcap S_n \rightsquigarrow \bigcirc S \ (Next\text{-}time \ rules)$$

where each S_i is an atomic state and $S \in \mathcal{S}_D$. These formulas are interpreted over a timed sequence of global states: $GS_0, \ldots, GS_n..$ $-S$ is used to express that a state is not true, $S_1 \sqcap S_2$ that two states are true at a same time, $S' \rightsquigarrow S$ represents causal influence of S' over S, and $S' \rightsquigarrow \bigcirc S$ represents delayed (next time) causal influence of S' over S. Same-time rules are required to be *stratified*, i.e., they can be ordered in such a way that the states in a rule are either independent or dependent on states which are heads of rules in previous levels of the stratification. See more formal details in [8].

We have implemented this logical notation with an equivalent translation using ASCII symbols: & represents logical 'and', # represents logical negation, occurs(ingr(s), t1:t2) represents the event of the system ingressing to a state s at the instant in between $t1$ and $t2$. *Same-time rules* are implemented through *State-State rules* and *Next-time rules* are implemented through *State-Event rules*. A *State-State rule* ssr(a => b) represents a being true causes state b to be true. A *State-Event rule* ser(a -> ingr(b)) represents a being true causes an ingression to state b being true.

Scenario (fall detection in bedroom): the system trying to aid an occupant by caring for her/his safety whilst allowing her/him to develop daily activities without intervention will face the challenge to decide whether it is worth requesting

Description Layers   Decision Layers

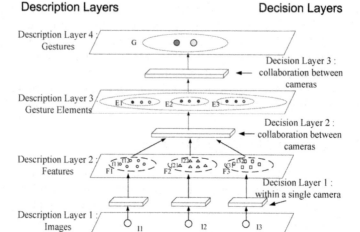

(a)

Original images and silhouettes   Elliptical representation
                                              and motion of body parts

(b)   (c)

Fig. 2. (a) The logical architecture of the vision system. R_i stands for images of camera i; F_i's are feature sets for R_i's; E_i's are posture element sets; and G is the set of possible postures. (b) Sample images from 2 cameras showing different postures and silhouettes. (c) Elliptical model representations with average motion vectors for the moving body parts.

the attention of the carer or not. Lets assume we have states: *inBed* representing the sensor detecting a person is in bed, *standing/moving/sitting/laying* representing the camera has captured an image which depicts the patient in one of those specific positions (see for example those in Figure 2(b)), *inbedTime* specifies the time window where the patient normally will go to bed, *mt20u* captures the passage of 20 units of time, *safe/undesirable* represents the belief that the person is in one of those situations, *carerVisits/carerCalls* represents that currently the carer has been required to take one of those actions, *carerGivesOK* is the input given by the carer through the system interface denoting s/he has checked or assisted the occupant, possibly via the bi-directional voice channel. We also consider a sequence of events and causal rules (only selected elements of the specification included):

```
states([inBed, standing, moving, sitting, laying, inbedTime, mt20u, safe,
        undesirable, carerVisits,carerCalls,carerGivesOK]).

is(inBed).     is(standing).     is(moving).     is(sitting).     is(laying).
is(inbedTime).  is(mt20u).        is(carerGivesOK).

occurs(ingr(sitting), 7:8).
occurs(ingr(mt20u), 28:29).
occurs(ingr(carerGivesOK), 31:32).

holdsAt(safe, 0).
holdsAt(inbedTime, 0).

ssr(# inBed & standing => safe).
ssr(# inBed & laying => # safe).
ssr(inBed & # inbedTime => undesirable).
ssr(inBed & sitting & inbedTime => safe).
ssr(inBed & sitting & inbedTime & mt20u => undesirable).

ssr(# safe => carerVisits).
ssr(undesirable => carerCalls).

ser(# safe & carerGivesOK -> ingr(safe)).
ser(undesirable & carerGivesOK -> ingr(# undesirable)).
```

We use a backward reasoning (and therefore more focused and efficient) version of the algorithm given in [8] which also allows us to obtain an explanation of the system's advice. This feature is important to understand the context of a warning. The causal connection in between the features of particular situations and potential unsafe conditions for the occupant is established through the rules. These relations give place to computational trees in which the root node is labelled by the head of a `ssr/ser` rule and the descendant nodes are labelled by the elements listed in the body of the rule. Thus the independent states always appear as leaves of the trees. We say a tree is *activated* at time t if at least one of its independent states, is known to hold at t; it is *fully activated* at t if *all* its independent states hold at t.

Our backward-reasoning algorithm requires us to find the set of trees supporting a given state S. This is achieved recursively: for each rule $B_1 \sqcap \cdots \sqcap B_n \rightsquigarrow S$ or $B_1 \sqcap \cdots \sqcap B_n \rightsquigarrow \bigcirc S$, we apply the process to find the trees supporting each of $B_1, \ldots, B_n$, and each combination of such trees is joined together to form a new tree. The base cases of the recursion are $S \in \mathcal{S}_I$ (their trees are single nodes). We shall denote by STT_S (STT_{-S}) the set of trees supporting S ($-S$). Whenever an event that can change the states supporting a causal tree is detected a reasoning process is triggered. To detect if a given hazardous State S is achieved at t, $Holds(S, t)$, our algorithm starts from t and works backwards trying to find the closest time to t at which a tree supporting either S or $-S$ becomes fully activated by the facts provided in the initial specification of the problem. If the first fully activated tree found supports S then the algorithm answers 'true' to $Holds(S, t)$; otherwise (the first fully activated tree found supports $-S$), it answers 'false'. If no tree supporting either S or $-S$ is found then the algorithm reaches time 0, so the truth value of S at t is determined by its truth value at 0 by persistence (that is, $STT_S \cup STT_{-S} \neq \emptyset$ is always true). The tree or fact used to answer the query is the causal explanation for the happening of S at t.

Given that the rules are required to be stratified and non-cyclic, it can be proven that the algorithm always finishes as the process always ends up at an intermediate time or eventually the answer is dependent on what happened at 0 (the initial time when the context currently analyzed was triggered).

In our Java/Prolog-based implementation `holds(State, GoalTime)` triggers a process to gather all the meaningful trees to prove or disprove that `State` is holding at `GoalTime`. These trees provide the alternative explanations for and against a position that `State` is holding at `GoalTime`. For example, if we want to know whether the state `undesirable` has been reached at say time 30, and if so a description of the context, we can query the system with `holds(undesirable,30)?` and the answer and explanation will be produced: [`inBed & sitting & inbedTime & mt20u => undesirable`]. A fact tells us the context assumes it is `inbedTime` at 0 and then actions are recorded through the sensors that the occupant goes `inBed` at 6, `sitting` at 8 and then more than 20 units of time elapses triggering a diagnosis of reaching a potentially `undesirable` situation, i.e., an indication the person may be unwell or has fallen asleep in an unhealthy position. If later on we want to know whether the state `undesirable` has been reverted few minutes later then `holds(undesirable,34)?` will be answered negatively explaining that the care giver took care of the situation (an undesirable state was reached at 29 and persisted until 31, after that point the caregiver's intervention allows the situation to be back to normal): [`undesirable & carerGivesOK -> ingr(# undesirable)`].

5   Related Work

Autominder [9] uses a scheduler and temporal constraints to guide elderly people through their daily tasks. One difference with our proposal beyond the specific hazard we consider is that they assume direct interaction with the technology on behalf of the occupant or, most likely, the carer. We provide a solution which does not rely on elderly people having to remember how to use technology and also reduces carer intervention to the essential, which is particularly important to enhance their quality of life.

Wearable sensors have been also used, see for example [10], to provide context-aware services to people on Smart Homes. In this work we explicitly want to depart from the assumption that the occupant has to pro-actively interact with devices in order to obtain a benefit. Although we recognize wearable devices are beneficial in some cases, we are focusing our efforts on exploring services that can be provided without depending of the assumption that a user is forced to wear a sensor or use a device (e.g., a PDA). We think that for the case of elderly and patients with dementia this assumption is unrealistic and does not take into account diminishing cognitive ability.

Other approaches either use full image processing [11] without caring for the need for intimacy or go to the other extreme refusing to use any image processing at all and therefore loosing their benefits. Here we support the thesis that image processing can be used in a way that does not diminish privacy and dignity. At the same time a simple outcome can be connected to an inferential unit capable of enhancing the service with advantages for occupants, carer and the health and social care system.

6   Conclusions

Given the importance of addressing ways to provide home care for elderly people, researchers have started to explore technological solutions to enhance health and social care provision in a way which complements existing services. The context of our research is the use of Smart Homes by a Health and Social Services Trust to enhance the quality of the care provision delivered to the elderly population and reduce hospital and nursing/residential home admissions. The solution proposed addresses the provision of intelligent care which is beneficial to the occupants, economically viable, and which can also improve the quality of life of the carers.

The problem of home care and in particular the detection of falls should not be underestimated. For example in one district of the UK, North Down and Ards in Northern Ireland, over 2600 people aged over 80 fall each year. This figure is much higher for people aged over 65 as 1 in 3 will fall per year. Many will eventually suffer a serious injury such as hip fracture which carries a six months mortality rate of 20%. In survivors, 50% are unable to return to independent living. Over 95% of hip fractures are falls related. The cost of individual hip fracture averages 29.000 Euros (45% acute care, 50% social care and long term hospitalisation, 5% drugs and follow up) [NHS-EW]. It has been estimated that in the UK falls in the elderly cost 1.4 billion Euros annually [12]. These figures are replicated in member states throughout Europe.

The solution described in this paper combines two key technologies: wireless sensor networks, enhanced by image capture and scene analysis and emergency voice channel, and inference logic which considers event sequence to trigger causal rules, for decision making. The proposed technique relies on vision technology but not in an intrusive way, to help detect falls. The occupant wears an unobtrusive badge for position sensing and voice communication, but is not required to interact directly with any technology.

It is recognized that image capture raises issues of privacy, and has ethical considerations, which may influence user acceptance. However statistics show that 90% of elderly people want to remain in their own home [13], and our solution strives to achieve a balance between optimal care and user acceptance.

With the use of wireless sensor networks, multiple image capture, and feature extraction algorithms, no human monitoring of images will be required under normal circumstances in practice. The images acquired have sufficient quality to facilitate segmentation and posture identification, but are treated within the camera node to obtain reconstructed body models or silhouetted images for reporting, hence allowing the proposed technique to offer a level of privacy and dignity to the user.

References

1. Nugent, C.D., Augusto, J.C. (eds.): In: Proceedings of the 4th International Conference on Smart Homes and Health Telematic (ICOST2006) vol. 19 of Assistive Technology Research, IOS Press, Belfast, UK (2006)
2. Shaw, F., Bond, J., Richardson, D., Dawsona, P., Steen, N., McKeith, I., Kenny, R.: Multifactorial intervention after a fall in older people with cognitive impairment and dementia presenting to the accident and emergency department: randomised controlled trial. British Medical Journal, pp. 326–373 (2003)
3. Gillespie, L., Gillespie, W., Robertson, M., Lamb, S., Cumming, R., Rowe, B.: Interventions for preventing falls in elderly people. Cochrane Database of Systematic Reviews 2003, vol. (4), pp. CD000340 (2003)
4. Tabar, A.M., Keshavarz, A., Aghajan, H.: Smart home care network using sensor fusion and distributed vision-based reasoning. In: Proc. of ACM Multimedia Workshop on VSSN (2006)
5. Keshavarz, A., Tabar, A.M., Aghajan, H.: Distributed vision-based reasoning for smart home care. In: Proc. of ACM SenSys Workshop on DSC (2006)
6. Wu, C., Aghajan, H.: Opportunistic feature fusion-based segmentation for human gesture analysis in vision networks. In: Proc. of IEEE SPS-DARTS (2007)
7. Wu, C., Aghajan, H.: Layered and collaborative gesture analysis in multi-camera networks. In: Proc. of IEEE ICASSP (2007)
8. Galton, A., Augusto, J.: Two approaches to event definition. In: Hameurlain, A., Cicchetti, R., Traunmüller, R. (eds.) Proceedings of 13th DEXA. LNCS, vol. 2453, pp. 547–556. Springer Verlag, Heidelberg (2002)
9. Pollack, M.E.: Intelligent Technology for an Aging Population: The Use of AI to Assist Elders with Cognitive Impairment. AI Magazine 26(2), 9–24 (2005)
10. Patterson, D., Fox, D., Kautz, H., Philipose, M.: Fine-grained activity recognition by aggregating abstract object usage. In: Proceedings of the IEEE Int. Symposium on Wearable Computers, Osaka, Japan, pp. 44–51. IEEE Press, New York (2005)

11. Natale, F.D., Mayora-Ibarra, O., Prisciandaro, L.: Interactive home assistant for supporting elderly citizens in emergency situations. In: Plomp, J., Tealdi, P. (eds.) Proceedings of Workshop on "AmI Technologies for WellBeing at Home", EU-SAI2004, Eindhoven, The Netherlands (2004)
12. Scuffham, P., Chaplin, S., Legood, R.: Incidence and costs of unintentional falls in older people in the united kingdom. Journal of Epidemiology and Community Health 57, 740–744 (2003)
13. Heaney, M.: Developing a telecare strategy in Northern Ireland. In: Proceedings of the Workshop on The role of e-Health and Asssistive Technologies in Healthcare, Ulster Hospital, UK (2006)

RNG-Based Scatternet Formation Algorithm for Small-Scale Ad-Hoc Networks

Chungho Cho and Gwanghyun Kim

Department of Information Communication Engineering
Gwangju University, 592-1 Jinwol-Dong, Nam-Gu, Gwangju, Korea, 503-703
Fax: 82-62-670-2183
chcho@gwangju.ac.kr, ghkim@gwangju.ac.kr

Abstract. This paper addresses a RNG based scatternet topology formation, self-healing, and self-routing path optimization for small-scale environment, called RNG-FHR(Relative Neighborhood Graph-scatternet Formation, self-Healing and Routing path-optimization) algorithm. Then, we also evaluated the performance of the algorithm using ns-2 and extensible Bluetooth simulator called blueware to show that even though RNG-FHR does not have superior performance than any other algorithms, it is simpler and more practical from the viewpoint of deploying the network in the distributed dynamic small-scale ad-hoc networks due to the exchange of fewer messages and the only dependency on local information. As a result, we realize that RNG-FHR is unlikely to be reasonable for deploying in large-scale environment, however, it surely appeals for practical implementation in small-scale environment.

Keywords: Bluetooth, Piconet, Scatternet, RNG, NS-2, Blueware.

1 Introduction

Bluetooth is a standard for short range, low-power wireless communication. It's MAC protocol has the function of constructing an ad hoc network without manual configuration or wired infrastructure. Bluetooth communication is based on TDD(Time Division Duplex) master-slave mechanism. A piconet is a group of nodes in which a master node controls the transmission of other slave nodes, which has the constraint that a piconet consists of one master and up to seven slaves. The basic piconet communication is done by alternating transmission slots, each odd slot of which being used by the slave, and each previous even slot of the odd slot is used by the master. Frequency hopping is used for multiple concurrent communication within radio range of each different piconet without performance degradation due to interference, which makes it possible high densities of communication devices to co-exit and independently communicate with other piconets, resulting in making the possibility of inter-working multiple piconets. Therefore, the frequency hopping principle could also create the new network concept, called scatternet, which is an ad hoc network

T. Okadome, T. Yamazaki, and M. Mokhtari (Eds.): ICOST 2007, LNCS 4541, pp. 206–216, 2007.

consisting of overlapping piconets. Self-healing in scatternet is to guarantee the join of new nodes and the removal of existing nodes due to mobility or failure/deactivation of nodes in dynamic environment. Routing path optimization is optimizing routing path to meet the requirements of the performance metrics such as overall network capacity.

In this paper, we present a practical scatternet formation, self-healing and optimized routing algorithm based on RNG algorithm, called RNG-FHR, which is much simpler, less message exchanging and more practical for implementation. Then, we evaluated the performance to show that it is practically possible for deploying in small scale distributed dynamic environments.

2 Preliminary Studies for Bluetooth Scatternet Formation Algorithms

We can describe some characteristic classification criteria for Bluetooth Scatternet Formation Algorithms[13]. The first one is a connectivity guarantee which means that each node is aware of all its neighbors within a communication range. The second one is a node degree which means the algorithms will be divided into those that guarantee the degree limitation for each created piconet or not. The third one is hop based criteria. The algorithms can be divided into those that are designed for the single-hop range and multi-hop range. In single-hop range each device is within communication range of any other device in the network, and it operates only when nodes are in radio visibility, that is, each device is within the radio transmission range of any other device by one hop, e.g., electronic devices in a laboratory, or laptops in a conference room[3][8]. However, in multi-hop range, some devices are not within transmission range of each other instead of being connected via other devices, and they can operate even though some pairs of nodes can't directly communicate with each other. The communication between them is done by passing the messages through the intermediate nodes[7][11][12].

The last one is knowledge degree of neighbor devices. The algorithms can be divided into those that each device is required to learn all its neighbors and learn some of its neighbors.

The link formation process in Bluetooth Scatternet consists of two processes: Inquiry and Page. The goal of the Inquiry process is for a master node to discover the existence of neighboring devices and to collect low level information which is primarily related to their native clocks. The goal of the Page process is to use the gathered information in Inquiry process to establish a bi-directional frequency hopping communication channel[3].

It produces connected scatternets by exploiting clustering schemes for ad hoc networks, which is primarily aimed at home and office environments. In multihop solutions, it requires the exact position information because they can influence on the formation of scatternet topology, each being composed of short range devices[9][10].

None of the previous schemes deals with RNG based small-scale dynamic environments where nodes may arbitrarily join or leave the network. However, our approach has the following characteristics[6].

- Nodes can arrive and depart at any time
- The topology can be healed if the node/link failure occurs
- The routing path can be optimized with the minimum cost

3 RNG Algorithm Concept

The key concept of RNG is adding links as and when they are discovered. Let |AB| denote the Euclidean distance between nodes A and B. RNG adds a link between two nodes, A and B in the logical topology if and only if A and B are in each others transmission range and |AB| <= max(|BC|, |AC|) for any other node C which is in A's and B's transmission ranges. After RNG has added AB, if a node C that violates the above condition is discovered, then RNG deletes AB.

The RNG has the following assumptions and characteristics.

- Each node is assumed to know its neighbors in the physical topology graph
- A node also knows an ordering among the Euclidean distances between its neighbors from power measurements and subsequent information exchange with its neighbors.
- Addition and/or deletion of a link do not affect any other link additions or deletions, and depend only on local information.
- There is no need to broadcast any information throughout. Thus, RNG exchanges fewer messages and is simpler than the other algorithms such as distributed MST[6].

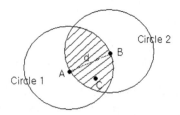

Fig. 1. As RNG algorithm, link AB is added if there is no other node in the shaded area. After link AB has been added, the link is deleted if node C is in the area.

4 RNG Based Scatternet Formation Algorithm

We assume that node 0 is controller which is interfaced to a bus and controls the communication between the other nodes. The other nodes(node1 to node 5) are master or slave node.

Phase 1 : Initiate each node
Piconet_No = [my_bd_addr], Node_Type = Free_Node.
If a node is controller, then Hops_to_Controller =0 else Hops_to_Controller = 1000.
Neighbor_Table(NT) = empty vector.
Format of NT is <Bd_addr, Sig_strength, Node_type, Roles, Ch_id, Piconet_no, Hops_to_controller>.
Bd_addr is the identifier of Bluetooth device.
Sig_strength is the signal strength from the the neighbor node.
Role is defined as one of As_Master/As_Slave/Not_Connected.
Ch_id is assigned to each link toward neighbor nodes if a node is connected to its neighbors. Otherwise, not assigned(Not_Assigned).
Neighbor_Piconet_Table(NPT) = empty vector.
Format of NPT is <Bridge_id, Remote_piconet_no>.
Bridge_id is the identifier of the device joining more than one piconet.
Remote_piconet_no is the identifier of the neighbor combined with the local piconet .

Phase 2 : Each node discovers neighbors and makes a physical link connection to one of the neighbors in baseband layer(or physical layer) according to Bluetooth specification.

a) Node A receives inquiry response from node C, meanwhile C creates a new entry for A in its NT(Neighbor_Table) based on the inquiry
b) Node A creates a new entry for C in its NT
c) Node A makes a physical link connection to C

Phase 3 : Once a physical link connection A(master node) -> C(slave node) is created in baseband layer, A and C exchange local information, and make a decision whether to create a logical connection based on the following rules. If it's rejected, disconnect the link connection

a) Master A sends RNG_REQ message to the slave C, formatted as <Bd_addr, Node_type, Piconet_no, Hops_to_controller>.
b) Upon receiving the message, the slave C makes a decision whether to accept this connection request. If accepting it, C sends back RNG_RESP message as a response and updates local configuration, otherwise disconnect it.
c) Upon receiving the responding RNG_RESP, the master A updates local information

Phase 4 : A and C exchange NT and make a decision whether to accept the new physical link based on two-hop neighbors information

a) A and C send their NT(defined as Neighbor_MSG) to each other
b) If (Both of A and C have connected E) then {

      Select the weakest link within the triangle;

If(the weakest link is A->E or C->E) then
        Disconnect the link, update local information and NT
else Disconnect A->C, update local information and NT }

Phase 5 : For A and C, if the Hops_to_controller is changed, A or C informs the neighbors, asking them to update Hops_to_controller. The neighbor also informs the change to their neighbors. In this way, all the nodes update the Hops_to_controller if they can find a shorter path.

a) If(local Hops_to_controller for A and C is changed) then

        Broadcast Hop_MSG to their connected neighbors, the format is <Bd_addr, Node_type, Piconet_no, Hops_to_controller>
b) On receiving Hop_MSG for all nodes
    If(local Hops_to_controller > Hops_to_controller in the Hop_MSG + 1) then {

        Update NT based on the received Hop_MSG;
        Set local Hops_to_controller = Hops_to_controller in the Hop_MSG + 1;
        Generate and broadcast Hop_MSG to all the neighbors; }

Phase 6 : Make a decision whether to stop discovery procedure
If all the node has (Hops_to_controller < 1000) then Stop discovery

5   RNG Based Scatter Formation Scenarios

Stage 1 : Node 0 is a controller and node (1~ 5) have any role of Master/Slave/Slave-slave bridge/Master-slave bridge depending on how to deploy the discovery. Assume that at first node 1 discovers node 2(see Fig. 2 and Fig. 3).

HTC(Hop_to_Controller) of Node 1 to Node 5 is 1000

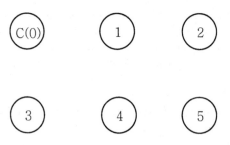

Fig. 2. Initial state

C: Controller, M : Master, S : Slave
HTC of Node 1 to Node 5 is 1000

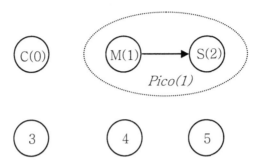

Fig. 3. Node 1 discovers node 2

Stage 2 : Assume that node 4 discovers node 2(see Fig. 4).

SS : Slave-Slave bridge

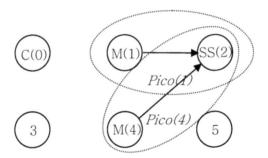

Fig. 4. Node 4 discovers node 2

Stage 3 : Assume that node 0 discovers node 3(see Fig. 5).

CM : Controller with Master node type
HTC of node 1 to node 5 except node 3 is 1000

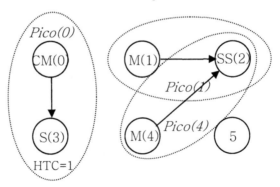

Fig. 5. Node 0 discovers node 3

Stage 4 : Assume that node 0 discovers node 4(see Fig. 6).

MS : Master-Slave bridge
HTC of node 3 and node 4 is 1
HTC of node 1, node 2 and node 5 is 1000

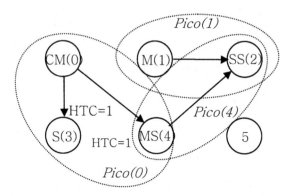

Fig. 6. Node 0 discovers node 4

Stage 5 : Assume that node 3 discovers node 4 (see Fig. 7).

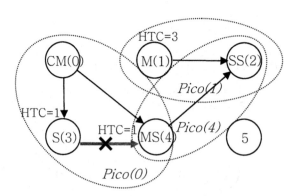

Fig. 7. Node 3 discovers node 4

Stage 6 : Assume that node 4 discovers node 1. RNG algorithms are applied to triangle (see Fig. 8).

Stage 7 : Assume that node 2 discovers node 5) (see Fig. 9).

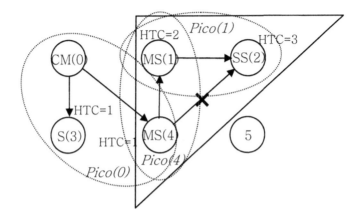

Fig. 8. Node 4 discovers node 1

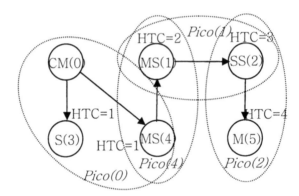

Fig. 9. Node 2 discovers node 5

6   Performance Evaluation

We evaluated our algorithm using Blueware[1][5], which is an extension to ns-2[2] and developed by Godfrey Tan. Blueware provides an extension architecture for various scatternet formation and link scheduling schemes[4]. We simulated to evaluate the performance of RNG-FHR with comparing with TSF algorithm[3]. In this section, we present and discuss the results on scatternet formation, link establishment, and converging to stable scatternet. In all the experiments, nodes are assigned to a random clock value, and every result data is the average of 30 independent trials. We are unable to compare exactly RNG-FHR's performance with TSF in evaluation of consuming time to stable state of topology because each algorithm has different simulation environments, particularly the schemes have different assumptions on the efficacy of consumption to accomplish the stable state of topology.

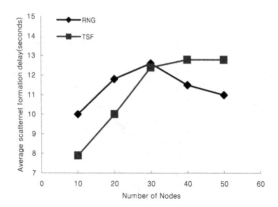

Fig. 10. Average scatternet formation delay

In RNG-FHR, the delays to form the scatternet for n nodes grow until 30 nodes, but go down if the number of nodes are more than 30 because the algorithm takes more messages to come to form the connected topology from the initial state, After the connected topology has been formed, the addition and deletion of a node or a link in RNG-FHR does not influence any other node/link and depends only on local information, thus, it exchanges fewer messages and simpler than the TSF algorithm. However, in TSF, the delay grows logarithmically with the size of the scatternet. We have an instinct that TSF achieves a logarithmic average scatternet formation delay. The data points for TSF are obtained from [3].

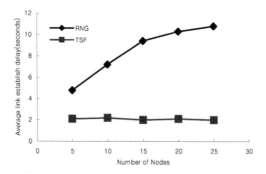

Fig. 11. Average link establish delay

In RNG-FHR, the average link establish delay means the time to take for a new node to connect to an existing network. The delay grows as the number of nodes increases, but the growing rate gets smaller and smaller as the number of nodes increases because each node exchanges messages and have enough information for neighbor nodes as time elapses. RNG-FHR spends much more time for a free node to get attached to an existing network. However, the delay in TSF is always less than 3 seconds because as the tree gets larger and larger, it gets a bit faster for a free node to get attached to a non-root node.

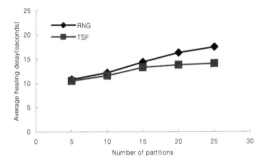

Fig. 12. Average healing delay

We evaluate the delay for healing network partitions when any node leaves. If any node leaves, the scatternet will be partitioned into smaller piconets. In this case, coordinating nodes try to connect each piconet during the healing the scatternet. The delay grows logarithmically with the number of network partitions. The delay for each algorithm are growing analogously for small number of partitons, but the growing rate of delay gets smaller for each algorithms as the number of partitions gets larger and larger because both of algorithm depend on only local information that each node maintains only about adjacent nodes. However, TSF may not able to heal all partitions when all the nodes are within radio proximity because TSF limits the tasks of discovering and merging partitions to coordinators and roots respectively.

7   Conclusion

RNG-SFR is RNG based   scatternet formation, self-healing, and self-routing path optimization which are done depending only on local information throughout the graph, so is more simpler than any other algorithm. It allows nodes to be added and/or deleted at any time, healing partitions and routing whenever they happen. Furthermore, it allows an added node to begin communication with neighbor nodes.

The performance are evaluated comparing to TSF. The algorithm is simpler and more practical than any other distributed algorithms from a point of view in deploying the distributed and dynamic small scale environment. Using ns-2 and extensible Bluetooth simulator called blueware, we simulated and evaluated comparing to TSF to show that the RNG-FHR has analogous latencies in scatternet formation and healing due to the exchange of fewer messages and the only dependency on local information. However, link establishment in RNG-FHR has larger latency than TSF. As a result, we notice that even though RNG-FHR is not superior to TSF in performance, it is more practical in implementing because of simplicity and local control.

As a result, RNG-FHR scatternet is unlikely that it will be deployed in large-scale ad hoc networks because it has less performance matrics and more complexity as the network size increases. However, it surely possible that the scatternet will be practically applied in deploying small-scale ad hoc networks.

References

[1] Blueware: Bluetooth simulator for ns http://nms.lcs.mit.edu/projects/blueware/ http://nms.csail.mit.edu/projects/blueware/software

[2] ns-2 Network Simulator http://www.isi.edu/vint/nsnam

[3] Tan, G., Miu, A., Balakrishnan, H., Guttag, J.: An Efficient Scatternet Formation Algorithm for Dynamic Environments. In: IASTED International Conference on Communications and Computer Network(CCN02), Cambridge, MA (November 2002)

[4] Tan, G., Guttag, J.: A Locally Coordinated Scatternet Scheduling Algorithm. In: The 27thAnnual IEEE Conference on Local Computer Networks(LCN), Tampa, FL (November 2002)

[5] Tan, G.: Blueware: Bluetooth Simulator for ns. MIT Technical Report, MIT-LCS-TR-866, Cambridge, MA (October 2002)

[6] Vergetis, E., Guerin, R., Sarkar, S., Rank, J.: Can Bluetooth Succeed as a Large-Scale Ad Hoc Networking Technology? In: IEEE Journal on Selected Areas in Communications. vol. 23(3) (March 2005)

[7] Li, X.-Y., Wang, Y., Wan, P.-J., Song, W.-Z., Frieder, O.: Localized Low-Weight Graph and Its Applications in Wireless Ad Hoc Networks. In: IEEE INFOCOM 2004 (2004)

[8] Salonidis, T., Bhagwat, P., Tassiulas, L., LaMaire, R.: Distributed Topology Construction of Bluetooth Personal Area Networks. In: IEEE INFOCOM 2001 (2001)

[9] Petrioli, C., Basagni, S., Chlamtac, I.: Configuring BlueStars: Multihop Scatternet Formation for Bluetooth Networks. In: IEEE Transactions on Computers, vol. 52(6) (June 2003)

[10] X.-Y., Li, Stojmenivic, I., Wang,Y.: Partial Delaunay Triangulation and Degree Limited Localized Bluetooth Scatternet Formation. In: IEEE Transactions on Parallel and Distributed Systems. vol. 15(4) (April 2004)

[11] Petrioli, C., Basagni, S., Chlamtac, M.: Configuring BlueStars: Multihop scatternet formation for Bluetooth networks. IEEE Transactions on Computers 52, 779–790 (2003)

[12] Li, X.-Y., Stojmenovic, I., Wang, Y.: Partial delaunay triangulation and degree limited localized Bluetooth scatternet formation. IEEE Transactions on Parallel Distributed Syste m 15, 350–361 (2004)

[13] Stojmenovic, I., Zaguia, N.: Bluetooth scatternet formation in ad hoc wireless networks, University of Ottawa (September 2004)

A Visual Editor to Support the Use of Temporal Logic for ADL Monitoring

A. Rugnone[1], F. Poli[1], E. Vicario[1], C. D. Nugent[2], E. Tamburini[3], and C. Paggetti[3]

[1] Department of Systems and Computer Science, University of Florence, Italy
[2] Faculty of Engineering, University of Ulster at Jorsanstown, Northern Ireland
[3] I+ srl, Florence, Italy

Abstract. The use of technology within the home environment has been established as an acceptable means to support independent living for elderly and disabled people. An area of particular interest within this domain relates to monitoring of Activities of Daily Living for those persons with a form of cognitive decline. In this area, specific tasks undertaken by the persons in the context of their normal day-to-day lives reveal a wealth of information to be used to customize their environment to improve their living experience. In our current work we investigate the development of models which can be used to represent, classify and monitor basic human behaviors and support observation and control of activities of daily living. In particular, in this paper we focus on the problem of automated recognition of sequences of events that may indicate critical conditions and unexpected behaviors requiring intervention and attention from caregivers. Our work is based on a formal framework developed with temporal logic used for the specification of critical sequences of patterns and a behavior checking engine for automated recognition. In addition we have also developed an approach to provide a means of interaction with user. A visual formalism for the specification of Linear Temporal Logic expressions reduces the barrier of technical complexity enabling the involvement of experts in the domain of healthcare services to interact with the system.

Kewwords: Model Checking, Temporal Logic, Activity Daily Living, Patient Behavior Models.

Introduction

As our elderly population continues to grow the need to develop intelligent environments to support independent living is ever increasing. Recent results have shown that the introduction of technological solutions within the home environment can have significant impact on the quality of life of the person. Examples of these solutions include basic motion sensors to activate lights, to safety devices, to prevent flooding in bathrooms, to elaborate wearable systems and to monitor and assess vital signs [1]. The common goal of all these solutions is to improve the living experience and quality of life for the person, customize their living environment and extend the period of time they can remain at home without the requirement for institutionalization. Although, on the whole, the introduction of technology can have positive effects, there are a number of potential concerns. Viable solutions for wide-scale uptake should not be expensive,

T. Okadome, T. Yamazaki, and M. Mokhtari (Eds.): ICOST 2007, LNCS 4541, pp. 217–225, 2007.

should be compatible or interoperable with other devices/solutions and should integrate with existing means for healthcare delivery. In addition, a realistic reimbursement scheme requires establishment for service realization.

In our current work, we have focused our efforts towards the monitoring and analysis of human behavior to support the control of Activities of Daily Living (*ADL*). Many efforts have recently been focused on ADL recognition: Augusto et al. [2] proposed a smart home framework centered on Event-Condition-Action (*ECA*) rules to capture events accuracy, applied to monitoring of activities of elderly people with cognitive impairment; Bauchet and Mayers [3] proposed a hierarchical model for the description of ADL; Bouchard et al. [4] proposed a non-quantitative logic approach to ADL recognition for Alzheimer's patients, based on lattice theory and action description logic.

In this paper, we propose a formal approach to critical behavior recognition in ADL analysis based on the use of temporal logic [5] and algorithms, usually applied in the context of model checking [6] [7], of Run Time Execution Monitoring (REM) [8] [9] and also in on-line systems intrusion detection [10].

Based on these developments, we describe a service model that may be deployed to exploit this solution. Then we describe, with a partial degree of completeness and formalism, the characteristics of the logic deployed and the model checking algorithm. With the formalisms in place, we exemplify the kind of behavior that the system can describe and devise the type of service organization which it can effectively support. We also focus our attention on the visual formalisms to support the specification of the logic formulae, to support the involvement of users in the development of such systems, especially those with expertise in the domain of home based healthcare provision.

1   Service Model

The service model is a means of defining the stakeholders, their responsibility, the links and the interaction between them, the devices and services and how the care will actually be delivered within the realms of home healthcare delivery. Within our current study we are initially focusing on the requirements of healthcare operators and experts within such a system. Healthcare operators and experts both have responsibility to define the necessary requirements to be used to monitor the patient. The expert user has the responsibility to define the expected behavior of the patient and the bounds of their deviation from normal conditions, while the operator has the responsibility to perform the necessary intervention when the patient is reporting a deviation from normal.

To provide an automated approach for service delivery the required system would require three main modules. These modules would involve:

1. A repository to store the representative behaviors (system rules) of the patient;
2. A monitoring platform to monitor multiple patients within their home environment [11];
3. A Behavior Checking Engine (BC Engine) acting as a model checker, which compares the specification of critical behavior against the observed sequence events.

Our work is based on the assumption that the subjects who are being monitored and there living environment is equipped with a suite of pervasive middleware sensors. The information provided by the sensors records low level events such as *"door is open"* or *"cooker is on"*. Using such information it is then possible to model the *ADL* as a set of observable variables that provide a sequence of conditions started with a initial state, formed by their initial values. The representation of what happens in the environment can be modeled by a sequence of states, differentiated by changes in measured values with events such as { *"Water added to the kettle"*, *"The kettle is full"*, *"Cooker has been turned on"*,... }. The operator monitors the single state of the end user via the monitoring platform. When a critical behavior occurs the Behavior Checking Engine automatically notifies the event via the monitoring platform. The format of the event notification issued by the Behavior Checking engine is dictated by the Behaviors Model Repository *(BMR)*.

The expert user has the responsibility to define the specification of the rules dictating the expected behavior of the patient. If, however, we wish to deploy complex formalisms to represent the underlying rules and model behavior it is necessary to provide a form of visual based rule editor which requires limited technical knowledge of the underlying notation. Based on the rules such an editor would produce, the model checking engine would have the ability to process them according to predefined formalisms on the fly. This approach makes possible the definition of a set of rules that may be deemed exportable and reusable all under the control of expert healthcare user.

The remainder of this paper focuses on the development of such an interface.

2 Behavior Checking

We propose the use of temporal logic to specify critical behaviors of persons within their home environment and a behavior checking algorithm to process the behaviors and hence identify their deviation from normal which implicitly represent situations of concern. In particular we propose the use of an adaptation of Past Linear Temporal Logic *(PLTL)* [12] to satisfy the assumptions made in Sect. 1 and also to deal with the constraints imposed by the application domain. The logic used is linear as we must accommodate for a sequence of states. In addition, we must accommodate for previous observations in addition to those at the present moment and the time elapsed between two events.

2.1 Temporal Logic : Syntax and Semantic

The language of temporal logic relies on describing conditions over paths (ρ), i.e. sequences of states. A formula defining rules on a path is called a *path formula*. Path formulae are built as a combination of atomic propositions on states, by means of Boolean connectors and temporal operators:

$$\gamma ::= p_s \mid \neg\gamma_1 \mid \gamma_1 \wedge \gamma_2 \mid \gamma_1 \mathbf{S}^\tau \gamma_2 \mid \mathbf{H}^\tau \gamma_1 \mid \mathbf{P}^\tau \gamma_2 \mid \mathbf{V}\gamma_2 \mid \mathbf{C}^\tau_{[min,max]}(\gamma_1)$$

where p_s is an atomic proposition on state. The temporal operators are $\mathbf{S}$ (Since), $\mathbf{H}$ (Historically), $\mathbf{P}$ (Past), $\mathbf{V}$ (Previous) and $\mathbf{C}$ (Count). By using an index τ we can specify that formulae built by these operators are constrained within a temporal interval $[-\tau, 0]$. It is also possible to omit τ from the writing of a temporal operator, thus implicitly setting it to ∞, and unbounding the operator time interest interval.

Meanings and uses of these operators are best described through the use of examples. Consider, in the first instance, a person within their home environment called Marco. A proposition which is not considered to be elementary could be represented by *"Marco doesn't add the water to the kettle if the kettle is full"*. We can decompose the phrase into two elementary propositions *"Marco doesn't add the water to the kettle"*, labelled with γ_1 , and *"The Kettle is full"*, labelled with γ_2. The initial statement can now be represented with the simple formula $\gamma_1 \wedge \gamma_2$. Another example is the following: *"Marco has been sleeping for the past 8 hours"* could be modelled as $H^{[8]}\gamma_1$, where γ_1 represents *"Marco sleeps"*. The logic can also be used to model more complex representations, for example consider the following; *"Marco is adding some flavour to his coffee and in the past he has used sugar"*. Based on the previous approaches we decompose the proposition into the following elementary components, *"Marco is adding some flavour to his coffee"*, γ_1, and "Marco has used sugar", γ_2, and hence the scenario can be redefined with the following formula $\gamma_1 \wedge P(\gamma_2)$. Consider, as a final example: *"The pasta has been in the pot since the gas was turned on 5 minutes ago"*. By adopting the same decomposition method, we can model the phrase with $\gamma_1 \, S \, \gamma_2$, where γ_1 is *"The pasta is in the pot"*, and γ_2 is *"the gas is on"*; taking ρ_4 a path $s_0...s_4$ as in 1. This means $\rho \models \gamma_1 \, S \, \gamma_2$, read ρ models $\gamma_1 \, S \gamma_2$, if it existed a sub-sequence of $\rho_3 = s_0...s_3$ that satisfies in a time minor of 5 minutes and each sub-sequence of $\rho_3 = s_0...s_3$ satisfies γ_1. With this scenario we have a trend on the time of formula as represented in the Fig. 1.

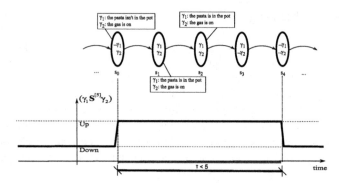

Fig. 1. The pasta is in the pot since the gas is on. In figure is represented the trend of formula on the sequence of states.

2.2   Behavior Checking Algorithms

The complexity of TL formulae tends to rapidly grow when applied to realistic behavioral pattern. 'Divide et Impera" is good practice to manage this complexity. To this scenario, we decompose instances of complex behavior into a set of simpler and complementary representations; however, doing so contributes to the increasing number of formulae required. In addition to this level of complexity and in accordance with the model to be created, the analysis of the sequence and duration of states require processing. This places the requirement for the model checking algorithm to satisfy rigid constraints relating to both time and space complexity and to provide a real time response.

To address these requirements, we propose an algorithm based on the recursive evolution of two variables, called Σ and T, whose calculation is done for each state notification. Σ means the validity of the formula to be checked, while T is a temporal evaluation on sequence of states. If Σ is true the formula satisfies the path which is currently being evaluated. Below, as an example we define the single step calculation, for the *Since* temporal operator previously introduced. Similar algorithms can be defined for the other operators.

$$\begin{cases} T_n = (s_n \models \gamma_2) \ ? \ \tau_{s_n} : T_{n-1} \\ \Sigma_n = (\,(s_n \models \gamma_2) \vee (\Sigma_{n-1} \wedge s_0 \models \gamma_1)\,) \wedge (\tau_{s_n} - T_n) \leq \tau \end{cases}$$

where s_n is the state presently observed, s_0 is the starting state of observation. τ_{s_n} is present time and $T_0 = 0$ and $\Sigma_0 = (\gamma_2 \models s_0)$ are the booting values of the sequence. It means, in the case of *Since* operator, that T_n stores the last time in which the condition γ_2 is satisfied on a state, while Σ_n envolves checking conditions both on state properties (the left hand of $\wedge$ operator), and on time (the right hand).

2.3 Visual Formalism

The Visual Editor is a tool based on simple visual formalisms that aims to make the process of specification of temporal logic based formula more *intuitively based* as opposed to *mathematical focused*. This approach facilitates the design of temporal logic expressions by Healthcare professionals, without them requiring an in depth knowledge of the underlying computational principles. The visual formalisms have been designed following a previously conducted user-centered usability engineering process [13]. An heuristic design process was applied to maximize consistency, i.e. to minimize the complexity of the visual metaphor mapping textual sentences into their respective visual representation.

The Basic visualization primitives associated with the semantic constructs of temporal progression, timing constraints, and propositional logic, can be combined into a set of visual re-writing rules, which associate each terminal token of PLTL with a concrete drawing. Matching the recursive organization of the PLTL syntax and semantics, these rules reduce the visualization of a formula to the recursive visualization of its subformulae and to the concrete drawing of graphic icons representing terminal symbols. While the structure of rules matches the PLTL syntax, the concrete drawings that these rules produce reflects the semantics of terminal symbols.

The PLTL expressivity combines three distinct fragments:

1. Basic Proposition;
2. Sequencing conditions;
3. Instantaneous conditions.

Following a commonly accepted standard, basic propositions are represented in textual form inside the frame of a test sheet icon (see Fig. 2(a)). Sequencing conditions expressed by temporal operators, underlie the basic concept of a sequence of temporal contexts through which the system progresses following the metaphor of timeline graphs. An example of such a sequence is visualized as a horizontal path line in Fig. 3.

For representation of instantaneous conditions captured through Boolean outputs, the process adopted suggests that operands should be arranged in a vertical direction (see Fig. 2). As an example, Figure 4 reports the visual representation for the textual formula of $C_{[3,inf]}(P^{[10min]}(A))$.

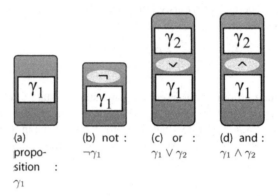

(a)   (b) not :   (c) or :   (d) and :
propo- $\neg \gamma_1$ $\gamma_1 \vee \gamma_2$ $\gamma_1 \wedge \gamma_2$
sition :

γ_1

Fig. 2. Graphical Logical operators. Figure 2(a) represents a proposition or an alias for a more complex formula (e.g. a predefined monitoring *macro*); figure 2(b) represents the negation of a subformula γ_1. 2(b), 2(c), 2(d) represent istantaneous conditions based on boolean operators.

3   An Example of Use

As a simple example of usage of the proposed approach, consider the scenario of a patient exhibiting a form of Confusional State Behavior. In this case, we consider two topics:

1. Repetitive Confusion : repetition of the same action in a relatively short time interval whilst not being considered as continuous;
2. Random Confusion : repetition of different actions in a relatively short time interval.

For instance, confusional behavior is a sequence like "A1, A2, A3, A1, A2, A3..." or a repetitive sequence like "A1, A1, A1...". Generally we can consider confusion as any cycle in the ADL model which is repeated too much in a short space of time. We can further progress this concept by defining *"too much"* repetitions as a quantity higher than 3 and *"a short space"* of time as a quantity less than 5 minutes. Although these are sharp decision boundaries in terms of temporal values and counter values, they are based only on experience in the domain, but are equally questionable. It is a generally accepted fact that this feature of temporal logic is not a good asset and can be considered as a major limitation of the approach. Nevertheless, these sharp decision boundaries are offset by good expressiveness and excellent flexibility to define system specifications, as we will examine. A cycle is a path that begins with a state, for example A, and also finishes with this state. We can model the confusion as $C_{[3,inf]}(P^{[10min]}(A))$ where A is any action or state, e.g. *"Add water to kettle"*, $P_{[10min]}(A)$ means at least 2 minutes ago action A is performed and, the last, that in the past $C_{[3,inf]}(P^{[10min]}(A))$ means

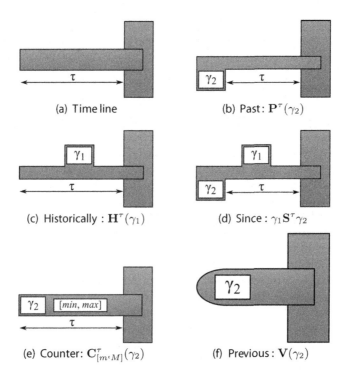

Fig. 3. Visual formalisms on time constraints and temporal operators. Figure 3(a) shows the basic concept of arranging graphical elements in the formalism: the horizontal dimension represents the temporal line, the horizontal rectangle represents a sequence of states evolving within temporal interval long τ, while the vertical rect represent the present. Figure 3(b) shows the formalism associated to the Past temporal operator, describing that at least once in the past, within a τ time from now, a condition γ_2 has been satisfied. Similarly, the other formalisms represent the remaining temporal operators: Historically (Fig. 3(c), requiring that a condition γ_1 holds continuously during the last τ time), Since (Fig. 3(d), requiring that since a condition γ_2 has been verified in the past, within a time τ from now, γ_1 holds continuosly), Counter (Fig. 3(e) requiring that a condition γ_2 has been satisfied a number of times between min and max during the last τ time) and Previous (Fig. 3(f) requiring γ_2 is satified that in the previous state).

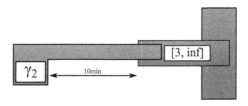

Fig. 4. Model of Confusion behavior. The formula $C_{[3,inf]}(P^{[10min]}(A))$ is expressed with use of visual formalism defined in Sect. 2.3.

that $P_{[10min]}(A)$ became true at least 3 times in the past. In similar way, an expert user can provide a set of template formula to describe a specific disease. This form of representation is modelled in Fig. 4 based on the visual formalism introduced in Fig. 4.

4   Conclusion

We have presented a formal model of a monitoring service based on temporal logic. We have deployed a behavior checking approach, relying on description of critical behaviors based on a temporal logic formalism, where behavior checking is considered as a meta-algorithm which can be programmed through the specification of a temporal logic formula rather than through direct programming. Through the addition of a visual layer on top of the temporal logic based formalisms, the use of such an approach by users with lower technical skills (relating to temporal logic) and higher domain knowledge, e.g. Healthcare professional, is made easier. This results in the only requirement from the user who is using the tool being one relating to specifying the needs of the task need and user requirements. This implicitly avoids the user requiring to know or indeed specify all possible permutations that patient can perform, but instead only to provide details on sequences of conditions that detect situations of concern, thus introducing a paradigm shift in the approach to comparison user's behavior against an expected task model. In fact, it offers the possibility to specify a set of *"abnormal"* behaviors without requiring the specification of the full set of *"normal"* ones, as required using operational user task models such as Automata, Petri Nets and Activity Diagrams.

References

1. Lymberis, A., de Rossi, D.: Wearable eHealth Systems for Personalised Health Management. In: State of the Art and Future Challenges, vol. 108, IOS Press, Amsterdam (2004)
2. Augusto, J.C., Nugent, C.D.: Designing Smart Homes, The Role of Artificial Intelligence. In: Augusto, J.C., Nugent, C.D. (eds.) Designing Smart Homes. LNCS, vol. 4008, Springer, Heidelberg (2006)
3. Bauchet, J., Mayers, A.: Modelisation of adls in its environment for cognitive assistance. In: Proc. of the 3rd International Conference on Smart Homes and Health Telematic (ICOST'05), Sherbrooke, Canada, pp. 3–10. IOS Press, Amsterdam (2005)
4. Bouchard, B., Giroux, S., Bouzouane, A.: A logical approach to adl recognition for alzheimer's patients. In: Proc. of the 4th International Conference on Smart Homes and Health Telematic (ICOST'06), pp. 1–8 (2006)
5. Manna, Z., Pnueli, A.: The Temporal Logic of Reactive and Concurrent Systems. Springer-Verlag, New York (1992)
6. Clarke, E.M., Emerson, E.A., Sistla, A.P.: Automatic verification of finite-state concurrent systems using temporal logic specifications. ACM Trans. Program. Lang. Syst. 8(2), 244–263 (1986)
7. Clarke, E.M., Grumberg, O., Hamaguchi, K.: Another look at ltl model checking. Formal Methods in System Design 10(1), 47–71 (1997)
8. Drusinsky, D., Shing, M.T.: Monitoring temporal logic specifications combined with time series constraints. J. UCS 9(11), 1261–1276 (2003)
9. Kim, M., Viswanathan, M., Kannan, S., Lee, I., Sokolsky, O.: Java-mac: A run-time assurance approach for java programs. Formal Methods in System Design 24(2), 129–155 (2004)

10. Naldurg, P., Sen, K., Thati, P.: A temporal logic based framework for intrusion detection. In: de Frutos-Escrig, D., Núñez, M. (eds.) FORTE. LNCS, vol. 3235, pp. 359–376. Springer, Heidelberg (2004)
11. Paggetti, C., Tamburini, E.: Remote management of integrated home care services: the dghome platform. In: Press, I. (ed.) From Smart Homes to Smart Care: Icost (2005), pp. 298–302 (2005)
12. Benedetti, M., Cimatti, A.: Bounded model checking for past ltl. In: Garavel, H., Hatcliff, J. (eds.) TACAS 2003. LNCS, vol. 2619, pp. 18–33. Springer, Heidelberg (2003)
13. Lusini, M., Vicario, E.: Engineering the usability of visual formalisms: a case study in real time logics. In: Catarci, T., Costabile, M.F., Santucci, G., Tarantino, L. (eds.) AVI, pp. 114–123. ACM Press, New York (1998)

Development of a Job Stress Evaluation Methodology Using Data Mining and RSM

Yonghee Lee[1], Sangmun Shin[2], Yongsun Choi[2], and Sang Do Lee[1]

[1] Department of Industrial Management Engineering, Dong-A University, Busan
South Korea
[2] Department of Systems Management Engineering, Inje University, Gimhae
South Korea
yonghee@donga.ac.kr, sshin@inje.ac.kr, yschoi@inje.ac.kr,
sdlee@dau.ac.kr

Abstract. Data mining (DM) has emerged as one of the key features of many applications on information system. While a number of data computing and analyzing method to conduct a survey analysis for a job stress evaluation represent a significant advance in the type of analytical tools currently available, there are limitations to its capability such as dimensionality associated with many survey questions and quality of information. In order to address these limitations on the capabilities of data computing and analyzing methods, we propose an advanced survey analysis procedure incorporating DM into a statistical analysis, which can reduce dimensionality of the large data set, and which may provide detailed statistical relationships among the factors and interesting responses by utilizing response surface methodology (RSM). The primary objective of this paper is to show how DM techniques can be effectively applied into a survey analysis related to a job stress evaluation by applying a correlation-based feature selection (CBFS) method. This CBFS method can evaluate the worth of a subset including input factors by considering the individual predictive ability of each factor along with the degree of redundancy between pairs of input factors. Our numerical example clearly shows that the proposed procedure can efficiently find significant factors related to the interesting response by reducing dimensionality.

Keywords: Job stress evaluation, Survey analysis, Data mining, Correlation-based feature selection (CBFS), Response surface methodology (RSM).

1 Introduction

The continuous improvement and application of the information system technology has become widely recognized by industry as critical in maintaining a competitive advantage in the marketplace. Despite the steady increase of computing power and speed, the complexity of today's many engineering analysis codes seems to keep pace with computing advances. Data mining (DM) has emerged as one of the key features of many applications on computer science, and has used as a means for predicting the future directions, extracting the hidden limitations, and the specifications of a

T. Okadome, T. Yamazaki, and M. Mokhtari (Eds.): ICOST 2007, LNCS 4541, pp. 226–237, 2007.
© Springer-Verlag Berlin Heidelberg 2007

product/process. The main purpose of DM is to define the nontrivial extraction of implicit, previously unknown, and potentially useful information from a large database [1]. In order to conduct this extraction successfully, DM uses computational techniques from statistics, machine learning and pattern recognition.

One of the most important encountered problems in a wide variety of industrial situations is to incorporate the voice of constituent members in an organization. Survey is an alternative method to consider various needs of the constituent members associated with workfare, environment and job stress. To conduct a survey analysis for job stress, it is very hard to find the specific influence of all factors affecting job stress because of dimensionality associated with a large number of factors (i.e., survey questions). However, most survey analysis methods for the job stress evaluation reported in literature may not provide comprehensive relationships among factors, and may just focus on frequency tests for input factors because of a large number of factors and their associated dimensionality. In order to address these problems, feature selection (FS) is known as an effective method for reducing dimensionality, removing irrelevant and redundant data, increasing mining accuracy, and improving comprehensibility of results among DM methods [2]. Consequently, FS has been a fertile field of research and development since 1970's and proven to be effective in removing irrelevant and redundant features, increasing efficiency in mining tasks, improving mining performance like predictive accuracy, and enhancing comprehensibility of results. The FS algorithm performs a search through the space of feature subsets [3]. In general, two categories of this algorithm have been proposed to solve FS problems. The first category is based on a filter approach that is independent of learning algorithms and serves as a filter to sieve the irrelevant factors. The second category is based on a wrapper approach, which uses an induction algorithm itself as part of a function evaluating factor subset. Because most filter methods are based on heuristic algorithms for general characteristics of the data rather than learning algorithms to evaluate the merit of factor subsets as wrapper methods do, filter methods are generally much faster, and has more practical capabilities to utilize high dimensionality than wrapper methods.

Most DM methods related to FS reported in literature may obtain a number of input factors associated with an interesting response factor without providing specific information, such as relationships among the input and response factors, statistical inferences, and analyses ([4], [5], [6] and [7]). Based on this awareness, Witten and Frank [7] suggested an alternative DM approach to a semiconductor-manufacturing problem in order to find significant factors. Su et al. [8] developed an integrated procedure combining DM and Taguchi methods. In order to consider a robust process design concept, Yi et al. [9] developed a robust data mining method for a wastewater treatment process.

While DM represents a significant advance in a type of analytical tools currently available, there are limitations in its capabilities to identify a causal relationship [10]. One of the limitations is that DM may not sufficiently identify a causal relationship although it can identify connections between responses and/or factors. Among survey analysis methods for job stress evaluation problems currently studied in the science and engineering community, researchers often identify the response surface methodology (RSM) as one of the most effective methodologies to reduce the inherent uncertainty associated with input factors and responses. However, most RSMs reported in literature may obtain the most favorable solution for a small number of given input factors without considering the reduction of dimensionality associated with many input factors (i.e., survey questions) and quality of information. Although a number of survey analysis methods for the job stress evaluation and their associated computer software packages are widely used, there is room for improvement in order to provide comprehensive relationships among a large number of factors and its associated response. In addition, few applications of DM methods to the survey analysis problem have conducted in the research community. To this end, we propose an enhanced survey analysis method for job stress evaluation problems incorporating a DM method into RSM. The primary objective of this paper is two-fold. First, we show how DM techniques can be effectively applied into the survey analysis for the job stress evaluation by applying a correlation-based feature selection (CBFS) method. This CBFS method can evaluate the worth of a subset including input factors by considering the individual predictive ability of each factor along with the degree of redundancy between pairs of input factors. When a large number of input factors are considered for the management of job stress, this CBFS method is far more effective than any other methods. After reducing the dimensionality associated with a large number of factors, we then conduct RSM in order to provide specific relationships among input factors and the interesting response. RSM is a statistical tool that is useful for modeling and analysis in situations where the response of interest is affected by several input factors. Our numerical example clearly shows that the proposed procedure can efficiently find significant factors related to the interesting response, and can effectively apply to a survey analysis for the job stress evaluation problem by reducing dimensionality of a large number of input factors. An overview of the proposed procedure integrating DM into RSM for the job stress evaluation problem can be shown in Figure 1.

2 Job Stress Evaluation Method

Job stress due to downsizing and restructuring the company leads to develop both mental disorder and physical illness of workers. It is strongly required that systemic studies to grasp the whole picture of current industrial situations for job stress and to clarify its associated risk factors needs to be performed. Although the levels of job stress and the detailed risk factors are often available in the real world industrial situations, risk assessment methods to consider job stress and risk factors can be categorized into two groups: macro-ergonomic and micro-ergonomic risk assessment tools. Limited research has been conducted with respect to development and testing of macro-risk assessment procedures [11]. However, the use of such techniques will become an important foundation on a health surveillance process due to the anticipated promulgation of ergonomic standards in various countries around the world. In order to assess ergonomic stress in workplaces, Lin et al. [11] developed a macro-ergonomic risk assessment tool. Macro-ergonomic approach implies the consideration of social, technological and physical works related to subsystems [12].

Although a number of researchers have regarded as psychosocial factors to evaluate job stress for the past 20 years, the exact etiology of job stress is currently unclear because the relationship between human and organization is complicate and organismic, and the patterns of job stress are muli-factorial ([13], [14]). Therefore, an integrated evaluation model for simultaneously considering psychosocial, physical and environmental factors related to job stress may essentially required in workplaces.

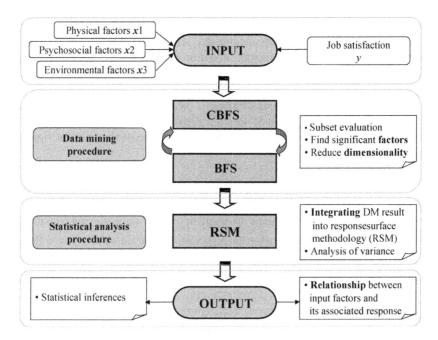

Fig. 1. The overview of the proposed integrated procedure for evaluating job stress: Three types of input factors are considered (physical, psychosocial, and environmental factors). The first stage represents a procedure of significant factor selection using the CBFS method, and the second stage implies a specific statistical analysis using RSM based on the results of the previous stage.

2.1 Classification of Job Stress Factors

In order to address job stress evaluation problems, many researchers have much attention to the classification of job stress factors. The majority of previous studies related to the classification of job stress factors have focused on psychosocial and few environmental factors without considering physical factors. For example, Job Contests Questionnaire (JCQ) [15] considered psychosocial factors, and Generic Job Stress Questionnaire (GJSQ) [16] suggested psychosocial and environmental factors. To this end, we propose a new survey method called an Integrated Job Stress Questionnaire (IJSQ) which incorporates not only all three classifications of job stress factors (i.e., psychosocial, environmental and physical factors) but also response factors related to job satisfaction for further effectively managing job stress levels as shown in Table 1.

3 Data Mining Method

3.1 Correlation-Based Feature Selection (CBFS) Method

CBFS is a filter algorithm that ranks subsets of input features according to a correlation based heuristic evaluation function. The bias of the evaluation function is toward subsets that contain a number of input factors, which are not only highly correlated with a specified response but also uncorrelated with each other ([10], [17] and [18]). Among input factors, irrelevant factors should be ignored because they may have low correlation with the given response. Although some selected factors are highly correlated with the specified response, redundant factors must be screened out because they are also highly correlated with one or more of these selected factors. The acceptance of a factor depends on the extent to which it predicts the response in areas of the instance space not already predicted by other factors. The evaluation function of the proposed subset is

$$EV_S = \frac{n\overline{\rho}_{FR}}{\sqrt{n + n(n-1)\overline{\rho}_{FF}}} \tag{1}$$

where EV_S, $\overline{\rho}_{FR}$, and $\overline{\rho}_{FF}$ represents the heuristic evaluation value of a factor subset S containing n factors, the mean of factor-response correlation $(F \in S)$, and the mean of factor-factor inter-correlation, respectively. $\sqrt{n + n(n-1)\overline{\rho}_{FF}}$ and $n\overline{\rho}_{FR}$ indicate the prediction of the response based on a set of factors and the redundancy among the factors. In order to measure the correlation between two factors or a factor and the response, an evaluation of a criterion called symmetrical uncertainty [19].

The symmetrical measure represents that the amount of information gained about Y after observing X is equal to the amount of information gained about X after observing Y. Symmetry is a desirable property for a measure of factor-factor inter-correlation or factor-response correlation. Unfortunately, information gain is not apt to factors with more values. In addition, $\overline{\rho}_{FR}$ and $\overline{\rho}_{FF}$ should be normalized to ensure they are comparable and have the same effect. Symmetrical uncertainty can minimize bias of information gain toward features with more values and normalize its value to the range [0, 1]. The coefficient of symmetrical uncertainty can be calculated by

$$\rho_{FF} = 2.0 \times \left[\frac{gain}{H(Y) + H(X)} \right] \tag{2}$$

where

$$H(Y) = -\sum_{y \in Y} p(y) \log_2(p(y))$$

$$H(Y \mid X) = -\sum_{x \in X} p(x) \sum_{y \in Y} p(y \mid x) \log_2(p(y \mid x))$$

$$gain = H(Y) - H(Y \mid X) = H(X) - H(X \mid Y) = H(Y) + H(X) - H(X, Y)$$

and where $H(Y)$, $p(y)$, $H(Y|X)$, and $gain$ represent the entropy of the specified response Y, the probability of y value, the conditional entropy of Y given X, and the information gain that is a symmetrical measure reflects additional information about Y given X, respectively.

Table 1. In order to consider job stress factors, the proposed IJSQ is classified by three categories, such as physical, psychosocial, and environmental factors. The response factor associated with job satisfaction is also incorporated.

Classification of factors	Number of factors	Job stress factors
Physical factors	7	Adjustable rests, simpler repetitive tasks, proper work surface, using vibrators, treatment with heavy things, needs to put force, uncomfortable posture
Psychosocial factors	50	Increased stringent performance standards, increased job repetition, increased job regimentation, increased job boredom, increased rigid work procedures with high production standards Task indication without consistency, the atmosphere of the office with vertical authority, disadvantage from sexual difference
Environmental factors	13	Task lighting, improper ventilation, summer humidity levels, winter humidity levels, summer temperature levels, winter temperature levels, too close of proximity to others creates feelings of crowding, noise disturbances, exposure to dangerous materials, danger of accident, insufficient workers, space, facilities and equipments, cleanness of workplace, overall work environment
Response	1	Job satisfaction

3.2 Best First Search (BFS) Algorithm

In much literature, finding a best subset is hardly achieved in many industrial situations by using an exhaustive enumeration method. In order to reduce the search spaces for evaluating the number of subsets, one of the most effective methods is the best first search (BFS) method that is a heuristic search method to implement CBFS algorithm [5]. This method is based on an advanced search strategy that allows backtracking along a search space path. If the path being explored begins to look less promising, the best first search can back-track to a more promising previous subset and continue searching from there. The procedure using the proposed BFS algorithm is given by the following steps:

```
Step 1. Begin with the OPEN list containing the start state, the
CLOSE list empty, and BEST← start state (put start state to BEST).

Step 2. Let a subset, θ = arg max EVₛ (subset), (get the state from
OPEN with the highest evaluation EVₛ).

Step 3. Remove s from OPEN and add to CLOSED.

Step 4. If EVₛ (θ ) ≥ EVₛ (BEST ), then BEST ← θ (put θ to BEST).

Step 5. For each next subset ξ of θ that is not in the OPEN or CLOSED
list, evaluate and add to OPEN.
```

Step 6. If BEST changed in the last set of expansions, go to step 2.

Step 7. Return BEST.

The evaluation function given in Eq. (1) is a fundamental element of CBFS to impose a specific ranking on factor subsets in the search spaces. In most cases, enumerating all possible factor subsets is astronomically time-consuming. In order to reduce the computational complexity, the BFS method is utilized to find a best subset. The BFS method can start with either no factor or all factors. The former search process moves forward through the search space adding a single factor into the result, and the latter search process moves backward through the search space deleting a single factor from the result. To prevent the BFS method from exploring the entire search space, a stopping criterion is imposed. The search process may terminate if five consecutive fully expanded subsets show no improvement over the current best subset.

4 Response Surface Methodology (RSM)

RSM is typically used to optimize the response by estimating an input-response functional form when the exact functional relationship is not known or is very complicated. For a comprehensive presentation of RSM, Box *et al.* [20] and Shin and Cho

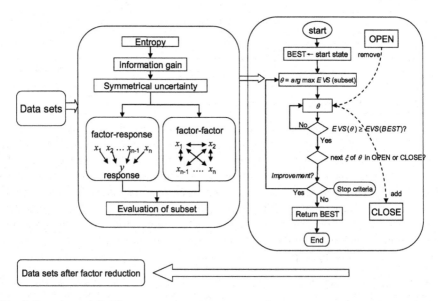

Fig. 2. Overview of CBFS method: The left side of this figure represents the subset evaluation procedure based on entropy, information gain, and symmetrical uncertainty by considering factor-response and factor-factor relationships. The right side of this figure illustrates the BFS algorithm in order to evaluate the subsets obtained by the subset evaluation procedure.

[21] provide insightful comments on the current status and future direction of RSM. Using this method, the response function for job stress evaluation is given by

$$\hat{y}(x) = \hat{\alpha}_0 + x^T a + x^T A x \qquad (3)$$

where

$$x = \begin{bmatrix} x_1 \\ x_2 \\ \vdots \\ x_k \end{bmatrix}, \quad a = \begin{bmatrix} \hat{\alpha}_1 \\ \hat{\alpha}_2 \\ \vdots \\ \hat{\alpha}_k \end{bmatrix}, \quad and \quad A = \begin{bmatrix} \hat{\alpha}_{11} & \hat{\alpha}_{12}/2 & \cdots & \hat{\alpha}_{1k}/2 \\ \hat{\alpha}_{12}/2 & \hat{\alpha}_{22} & \cdots & \hat{\alpha}_{2k}/2 \\ \vdots & \vdots & \ddots & \vdots \\ \hat{\alpha}_{1k}/2 & \hat{\alpha}_{2k}/2 & \cdots & \hat{\alpha}_{kk} \end{bmatrix},$$

where x_i terms are control factors, and the estimate of the α's in the function is estimated regression coefficients of the second-order fitted response function.

5 Numerical Example

In oder to evaluate job stress, a survey was conducted with consideration of the three factors categorized in Table 1. The survey was performed by a number of workers who are occupied in a manufacturing company. The company consiting appropriately 200 workers produces marine switchboards, control consoles and thrusters. Manufacturing processes include assembly (over 85%), welding and painting. This numerical example consists three parts of analysis results, such as a comparative study with consideration of previous methods in terms of overall basic statistics, significant factor selections using data mining, and statistical analyses using RSM.

As shown in Table 2, the results of the proposed IJSQ and previous methods provide similar values in mean and standard deviation. As shown in Table 3, DM results obtained by Weka software package [22] indicate that seven factors are significant to evaluate job stress. Among physical factors, 'improper work surface' was selected. Four factors including 'decreased superior support', 'decreased peer cohesion', 'increased job future uncertainty', and 'not up to the expectation of occupation' are selected in the psychosocial category. Among environmental factors, two significant factors were selected, such as 'danger of accident' and 'insufficient workers, space, facilities and equipments for tasks'.

Table 2. A comparative study is conducted using basic statistics (mean and variance) associated with the degree of job stress among the proposed survey method (IJSQ) and other methods (JCQ and GJSQ) to verification purposes.

Survey methods	Degree of job stress		Job satisfaction	
	Mean	Standard Deviation	Mean	Standard Deviation
IJSQ	2.48	0.28	2.20	0.33
JCQ	2.47	0.24	.	.
GJSQ	2.43	0.25	.	.

Based on the FS results, RSM was performed using MINITAB software package [23]. As shown in Table 4, x_1, x_2, x_3, x_5, x_6 , and x_7 among main effects are significantly affect to the response based on significant level 0.05 for the individual t-test. $x_1* x_5$, $x_3* x_4$, and, $x_3* x_5$ among interaction effects are also significant. The global F-test indicates that the regression is significant.

Table 3. Data mining results: In order to evaluate job stress, job satisfaction is regareded as the interesting response factor. Seven input factors (one physical, four psychosocial, and two environmental factors) are selected from the proposed DA method.

		The response factor	Job satisfaction	
		Merit of best subset	0.85	
Selected Evaluator		Selected factors	Physical factors	Improper work surface
			Psychosocial factors	Decreased superior support
				Decreased peer cohesion
				Increased job future uncertainty
				Not up to the expectation of occupation
			Environmental factors	Danger of accident
				Insufficient workers, space facilities and equipments for tasks
Search method		Search method	Best First	
		Search Direction	Forward	
		Total number of subsets evaluated	779	

Using Eqs. 3, the fitted response function can be estimated as follows:

$$\hat{y}(x) = 2.1851 + x^T a + x^T A x$$

where

$$x = \begin{bmatrix} x_1 \\ x_2 \\ x_3 \\ x_4 \\ x_5 \\ x_6 \\ x_7 \end{bmatrix} \quad a = \begin{bmatrix} 0.1170 \\ 0.1099 \\ 0.2529 \\ 0.0086 \\ 0.1576 \\ 0.1542 \\ 0.0917 \end{bmatrix}, \quad and \ A = \begin{bmatrix} 0.0228 & 0.0038 & 0.0407 & -0.0268 & 0.0662 & 0.0347 & 0.0016 \\ 0.0038 & 0.0335 & -0.0049 & -0.0350 & -0.0091 & 0.0398 & -0.0045 \\ 0.0407 & -0.0049 & 0.0585 & 0.0893 & 0.0686 & -0.0069 & -0.0040 \\ -0.0268 & -0.0350 & 0.0893 & -0.0403 & 0.0154 & 0.0018 & -0.0083 \\ 0.0662 & -0.0091 & 0.0686 & 0.0154 & -0.0141 & 0.0404 & 0.0154 \\ 0.0347 & 0.0398 & -0.0069 & 0.0018 & 0.0404 & 0.0216 & 0.0321 \\ 0.0016 & -0.0045 & -0.0040 & -0.0083 & 0.0154 & 0.0321 & 0.0434 \end{bmatrix}$$

In terms of model fitness, the response mode is adequate to utilize a response function since it has 89.1% R-sq.

Table 4. RSM Results: x_1, x_2, x_3, x_4, x_5, x_6 , and x_7 represent decreased superior support; danger of accident; decreased peer cohesion; increased job future uncertainty; insufficient workers, space facilities and equipments for tasks; not up to the expectation of occupation; improper work surface, respectively.

Predictors	Coef	SE Coef	t	p
Constant	2.18509	0.02972	73.520	0.000
x_1	**0.11702**	**0.04156**	**2.816**	**0.006**
x_2	**0.10985**	**0.03293**	**3.336**	**0.001**
x_3	**0.25291**	**0.07615**	**3.321**	**0.001**
x_4	0.00862	0.03484	0.247	0.805
x_5	**0.15759**	**0.05580**	**2.824**	**0.006**
x_6	**0.15417**	**0.03885**	**3.968**	**0.000**
x_7	**0.09166**	**0.04269**	**2.147**	**0.036**
$x_1 * x_1$	0.02283	0.02943	0.776	0.441
$x_2 * x_2$	0.03349	0.03475	0.964	0.339
$x_3 * x_3$	0.05847	0.03389	1.725	0.089
$x_4 * x_4$	-0.04031	0.03529	-1.142	0.258
$x_5 * x_5$	-0.01412	0.05275	-0.268	0.790
$x_6 * x_6$	0.02161	0.03507	0.616	0.540
$x_7 * x_7$	0.04344	0.03395	1.279	0.205
$x_1 * x_2$	0.00756	0.03494	0.216	0.829
$x_1 * x_3$	0.08132	0.05989	1.358	0.179
$x_1 * x_4$	-0.05355	0.04975	-1.076	0.286
$x_1 * x_5$	**0.13233**	**0.04469**	**2.961**	**0.004**
$x_1 * x_6$	0.06932	0.04976	1.393	0.168
$x_1 * x_7$	0.00319	0.04007	0.080	0.937
$x_2 * x_3$	-0.00988	0.06814	-0.145	0.885
$x_2 * x_4$	-0.07005	0.04078	-1.718	0.091
$x_2 * x_5$	-0.01824	0.05818	-0.313	0.755
$x_2 * x_6$	0.07954	0.04442	1.790	0.078
$x_2 * x_7$	-0.00890	0.04679	-0.190	0.850
$x_3 * x_4$	**0.17862**	**0.07760**	**2.302**	**0.025**
$x_3 * x_5$	**0.13727**	**0.06072**	**2.260**	**0.027**
$x_3 * x_6$	-0.01371	0.09742	-0.141	0.889
$x_3 * x_7$	-0.00790	0.05914	-0.134	0.894
$x_4 * x_5$	0.03069	0.05302	0.579	0.565
$x_4 * x_6$	0.00358	0.04255	0.084	0.933
$x_4 * x_7$	-0.01661	0.04729	-0.351	0.727
$x_5 * x_6$	0.08079	0.08203	0.985	0.328
$x_5 * x_7$	0.03084	0.05257	0.587	0.560
$x_6 * x_7$	0.06419	0.04388	1.463	0.148

S = 0.09352   R-Sq = 89.1%   R-Sq(adj) = 83.2%

Analysis of Variance

Source	DF	SS	MS	F	p
Regression	35	4.59021	0.131149	14.99	0.000
Linear	7	3.91662	0.226990	25.95	0.000
Square	7	0.17734	0.008160	0.93	0.488
Interaction	21	0.49626	0.023631	2.70	0.001
Residual Error	64	0.55979	0.008747		
Total	99	5.15000			

6 Conclusion

In this paper, we first developed a new job stress evaluation model by integrating a DM method to find significant factors into the analysis of IJSQ. Based on the results of the DM method, we then found important factors affecting job stress among a large set of data. The CBFS method in its pure form is exhaustive, but the use of a stopping criterion makes the probability of searching the whole data set quickly. Finally, we showed that the proposed DM method could efficiently find significant factors by reducing dimensionality. Using DM results, RSM was conducted to provide statistical relationships among input factors and their associated response. Survey data from manufacturing company were used to conduct the numerical example. The proposed method effectively demonstrated significant factors related to the interesting response by providing detailed statistical inferences in this particular example. In order to manage quality of raw data, scientific outlier tests using expectation maximization (EM) algorithm can be a possible further research opportunity.

References

1. Frawley, W., Piatetsky-Shapiro, G., Matheus, C.: Knowledge Discovery in Databases: An Overview. AI Magazine, Fall, pp. 213–228 (1992)
2. Yu, L., Liu, H.: Feature Selection for High-Dimensional Data: A Fast Correlation-Based Filter Solution. In: The Proceedings of the 20th International Conference on Machine Leaning (ICML-03). Washington, DC, pp. 856–863 (2003)
3. Allen, D.: The Relationship between Variable Selection and Data Augmentation and a Method for Prediction. Technometrics 16, 125–127 (1974)
4. Press, W.H., Flannery, B.P., Teukolsky, S.A., Vetterling, W.T.: Numerical Recipes in C. Cambridge University Press, Cambridge, UK (1988)
5. Quinlan, R.R.: Induction of Decision Trees. Machine Learning, vol. 1(1), Hingham, MA, pp. 81–106 (1986)
6. Gardner, M., Bieker, J.: Data Mining Solves Tough Semiconductor Manufacturing Problems. In: Conference on Knowledge Discovery in Data Proceedings of the Sixth ACM SIGKDD International Conference on Knowledge Discovery and Data Mining. New York, pp. 376–383 (2000)
7. Witten, I.W.H., Frank, E.: Data Mining: Practical Machine Learning Tools and Techniques, 2nd edn. Morgan Kaufmann, San Francisco (2005)
8. Su, C.T., Chen, M.C., Chan, H.L: Applying Neural Network and Scatter Search to Optimize Parameter Design with Dynamic Characteristics. Journal of the Operational Research Society 56, 1132–1140 (2005)
9. Yi, G., Choi, M.G., Choi, Y.S., Shin, S.M.: Development of a Data Mining Methodology Using Robust Design. WSEAS Transactions on Computers 5(5), 852–857 (2006)
10. Seifert, J.W.: Data Mining: An Overview. CRS Report RL31798 (2004)
11. Lin, G., Ash, G., Doran, C., Kevin, H.: Macro-ergonomic Risk Assessment in Nuclear Remediation Industry. Applied Ergonomics, pp. 241–254 (1996)
12. O'Neill Michael, J.: Ergonomic Design for Organizational Effectiveness. Lewis Publishers (1998)

13. National Research Council and the Institute of Medicine: Musculoskeletal Disorders and the Workplace: Low Back and Upper Extremities. Panel on Musculoskeletal Disorders and the Workplace. Commission on Behavioral and Social Sciences and Education. National Academy Press, Washington, DC (2001)
14. Deveraux, J.J., Vlachonikolis, I.G., Buckle, P.W.: Epidemiological Study to Investigate Potential Interaction between Physical and Psychosocial Factors at Work that may Increase the Risk of Symptoms of Musculoskeletal Disorder of the Neck and Upper Limb. Occupational and Environmental Medicine 59(4), 269–277 (2002)
15. Karasek, R., Gordon, G., Pietrokovsky, C., Rrese, M., Pieper, C., Schwartz, J., Fry, L., Schirer, D.: Job Content Questionnaire: Questionnaire and User's Guide. Lowell, University of Massachusetts (1985)
16. Hurrell, J.J., McLaney, M.A.: Exposure to Job Stress – a New Psychometric Instrument. Scand. J. Work Environ. Health 14, 27–28 (1988)
17. Hall, M.A.: Correlation-based Feature Selection for Machine Learning. Ph.D diss. Waikato University. Department of Computer Science. Hamilton, New Zealand (1998)
18. Xu, Q., Kamel, M., Salama, M.M.A.: Significance Test for Feature Subset Selection on Image Recognition. In: International Conference of Image Analysis and Recognition (ICIAR-04), porto, Portugal. LNCS, vol. 3211, pp. 244–252. Springer, Heidelberg (2004)
19. Langley, P.: Selection of Relevant Features in Machine Learning. In: Proceedings of the AAAI Fall Symposium on Relevance, pp. 140–144. AAAI Press, Stanford (1994)
20. Box, G.E.P., Bisgaard, S., Fung, C.: An Explanation and Critique of Taguchi's Contributions to Quality Engineering. International Journal of Reliability Management 4, 123–131 (1988)
21. Shin, S., Cho, B.R.: Bias-specified robust design optimization and its analytical solutions. Computer & Industrial Engineering 48, 129–140 (2005)
22. Merz, C.J., Murphy, P.M.: UCI Repository of Machine Learning Database. http:// www. ics.uci.edu/ mlearn/MLrepository.html
23. MINITAB: http://www.minitab.com/

Supporting Impromptu Service Discovery and Access in Heterogeneous Assistive Environments

Daqing Zhang, Brian Lim, Manli Zhu, and Song Zheng

Institute for Infocomm Research
21 Heng Mui Keng Terrace, Singapore 119613
{daqing,yllim,mlzhu,szheng}@i2r.a-star.edu.sg

Abstract. Wireless hotspots are permeating the globe bringing interesting services and spontaneous connectivity to mobile users. In order to enable the elderly and disabled to be fully integrated into the society, it's of paramount importance to build a pervasive assistive environment where assistive services can be automatically discovered and easily accessed with the device-to-hand across the physical spaces. In this paper, we propose a framework that can support impromptu service discovery and context-aware service access with mobile devices in heterogeneous assistive environments. Different from the existing approaches, the framework requires no specialized hardware or software installation in mobile client devices, it can automatically generate personalized user interfaces based on context such as user preference and device capability. To demonstrate the effectiveness of the framework, we prototyped a set of assistive services in public spaces like shopping malls, leveraging the OSGi-based platform accessible from any WLAN enabled mobile devices.

1 Introduction

The rapid growth of the world's older and dependent population calls for assistive services in various living spaces such as home, office, hospital, shopping mall, museum, etc.. Mobile and pervasive technologies are opening new windows not only for ordinary people, but also for elderly and dependent people due to physical/cognitive restriction. As wireless hotspots are permeating the globe bringing interesting services and spontaneous connectivity to mobile users, one trend is to gradually transform various physical environments into *"assistive smart spaces"*, where dependent people can be supported across all the heterogeneous environments and be fully integrated into the society with better quality of life and independence.

In order to provide services in various assistive environments, impromptu service discovery and access with heterogeneous mobile devices become critical. There are two possible ways to make the services accessible to mobile users. One way is to get connected to a global service provider and access the location-based services through the service provider. In this case, the service provider needs to aggregate all the services in the smart spaces and have the indoor/outdoor location of the user. Due to the evolutionary nature of the autonomous smart spaces, it is unlikely to have such a powerful service provider in the near future. The other way, which we are in favor of,

T. Okadome, T. Yamazaki, and M. Mokhtari (Eds.): ICOST 2007, LNCS 4541, pp. 238–246, 2007.

is to enable the mobile user interact with individual smart space directly when the mobile user physically enters the space, that is, the mobile user can automatically discover and access the services provided by the smart spaces. In the latter case, the mobile user needs to get connected directly to the individual smart space using short range wireless connectivity.

There are two key challenges to achieve impromptu service discovery and access. The first challenge is how to automatically discover the relevant services upon entering the smart space; the second challenge is how to generate a personalized user interface for the specific mobile device based on user's preference and device capability. To address the first challenge, various software [1][2][3] and hardware [4][5][6] "plug-in" methods have proposed. The key idea was to embed specialized software or hardware components in both the mobile device and the environment which follow the same service discovery protocol, however, there is still no dominant solution that was standardized and accepted by the whole community. To address the second challenge, various device-independent user interface generation mechanisms [7][8][9][3] have proposed. The key idea was to separate the definition of a user interface from that of the underlying service, however, the user profile and interaction style were not well studied for impromptu generation of user interfaces for dependent people.

In this paper, we aim to design a new framework for dynamic service discovery and access in heterogeneous assistive environments, using a wide spectrum of WLAN enabled mobile devices. The proposed framework is designed to achieve the goal of impromptu service discovery and personalized device-independent service access. While the framework requires "minimum" set-up on the mobile devices, it is capable of generating appropriate user interfaces in mobile devices, according to the user preference and device context.

The rest of the paper is organized as follows: Section 2 starts with a use scenario for assisting a disabled person in a shopping mall environment, from the use case six system requirements are identified for supporting impromptu service discovery and access in heterogeneous assistive environments. After presenting the overall system design for impromptu service discovery and access framework in Section 3, an automatic service user interface generation mechanism according to the user context and device capability is described in Section 4. Then Section 5 gives the implementation details about the service framework and some assistive services in a shopping mall environment. Finally, some concluding remarks are made.

2 Use Case Scenario

Before elaborating technical details on the system, we would like to provide a scenario motivating the assistive services which need impromptu discovery and access in smart environments. From the scenario, system requirements can be derived.

Bob, a person with disability on wheelchair, would like to buy some movie DVDs in a new shopping mall. Upon entering the mall, he is presented a navigation service on his wireless PDA so that he finds that one DVD shop is located in third floor. When approaching the nearest lift, he finds a lift control service in his PDA as he is not able to push the button in the lift. He accesses the lift control service and see a user interface with big fonts as it's difficult for him to touch small icons. He selects level 3 and finds the DVD shop. In the shop, he is presented with 2 services: one is the DVD finder which helps him locate the DVD in a specific shelf, the other is a movie recommendation service that recommends movie DVDs based on his preference.

Apparently, in the above use scenario, impromptu and context-aware service discovery with mobile devices is a first requirement; the automatic generation of user interface of the selected services is also needed, based on the user's preference, physical constraints as well as the device capability. Concretely, the following requirements need to be satisfied in the mobile device and environment:

- R1: Minimum assumption on mobile devices: One of the first ideal features for impromptu service access is that the mobile device doesn't need to install any specialized software or hardware in order to interact with the services in a certain environment. The "minimum" assumption means leveraging general purpose software/hardware set-up already deployed in mobile devices.
- R2: Automatic discovery of services in heterogeneous spaces: Another ideal feature for impromptu service access with mobile devices is the automatic discovery of the relevant services in the smart environments.
- R3: Heterogeneous service aggregation: As the devices and services are heterogeneous, they also vary greatly in different environments. There should be an effective mechanism and platform that facilitates the management of both static and dynamic services associated with each smart space.
- R4: User and environment context consideration: In order to present the right services to the right person in the right form, user and environment context should be taken into account for service discovery and access in each smart space.
- R5: User interface generation according to user context and device capability: As the users and access devices vary, impromptu and personalized user interface generation based on user preference and device capability should be supported.
- R6: Secure service access and information exchange: When the mobile users access certain services, user inputs and the output of the services need to be exchanged in a secure manner. Privacy should be guaranteed as well.

3 System Architecture for Impromptu Service Discovery

To address the above-mentioned requirements, we propose and implement a impromptu service discovery and access framework which enables the services in heterogeneous environments accessible to any WLAN enabled mobile device, according to individual user's preference, situation and device capability. The system is designed corresponding to the six identified requirements as follows:

- R1: In order to communicate locally with the smart space, we assume that the mobile devices should have at least the built in WiFi chipset to allow wireless connectivity and a web browser to access the web server hosting the services of the smart spaces. The requirement on mobile devices is "minimum" as it doesn't rely on any specialized software or hardware. Both the WiFi capability and the web browser are general purpose set-up popularly found in many mobile devices.

- R2 & R3: Following the minimum assumption on mobile devices, it requests that the services are associated with a specific "smart space" like a shopping mall, a lift, or a DVD shop. When the mobile devices are detected within a certain smart space, the services can be automatically discovered and presented in the mobile devices. The automatic service discovery is enabled by the captive portal and service portal mechanism as shown in Fig. 1. When the mobile device connects to a wireless hot-spot, the browser is detected by the Captive Portal (1), and forwarded to the Service Portal (2). The user can then browse the available services (3), and invoke a service through his mobile device (4). We adopted the open standard-based service platform such as OSGi [10] to aggregate various services and provide the service portal in a specific smart space. OSGi is known as a service platform which already supports various devices and services such as UPnP, Jini, HAVi, X-10, etc.

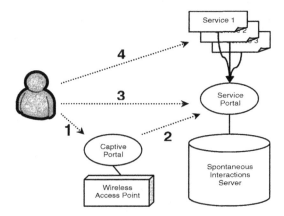

Fig. 1. Service discovery through captive and service portal mechanism

- R4: The relevant contexts that might facilitate service discovery/access include the user preference, user identity, location, orientation, and the device context such as screen size and input/output constraints. The "Semantic Space" [11] has been proposed and implemented to manage the context about users, devices and the environment. The function of the context manager is to acquire, process and provide contexts to specific context-aware services in smart spaces. Some examples of the context are: "who is approaching the lift ?", "what's user's preference about watching movie ?" , "what is the screen size of the mobile device ?".
- R5: Separating the user interface from the service functionality is the key for the generation of user interface to different targeted platforms. The user constraints and device constraints have given new impetus to the need for personalized user interface generation. The detailed user interface generation mechanism is given in Section 4.
- R6: As the web browser is the user interface shown in the mobile device and connected with the services in the smart space, thus certain security mechanisms and information exchange profiles need to be defined and agreed upon. In the system design, the user authentication and information exchange profile are associated

with individual service. When a service gets accessed, it will invoke the associated authentication and information exchange process. The W3C Composite Capabilities/Preferences Profile (CC/PP) [12] is recommended to exchange information between the personal device and the environment services.

The overall system architecture is show in Fig. 2 where each smart space has a impromptu service discovery and access framework embedded as a server. When the dependent people move from one space to another with a WLAN enabled mobile device, his mobile device's web browser will be directed by the captive portal in the framework to the service portal of the new smart space. The middleware core refers to the OSGi service platform, which aggregates the heterogeneous devices and services (UPnP, Jini, HAVi, Web Services, etc.) in each space. The context manager is the context middleware "Semantic Space", which facilitates service discovery and access with mobile devices. The services refer to the aggregated services offered by the devices and software sources in each space. Based on the user preference and device capability, the UI generator automatically generates the appropriate user interface for the service accessed on the device-to-hand.

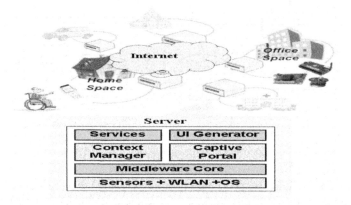

Fig. 2. Impromptu service discovery and access framework for multiple spaces

4   Automatic Generation of Personalized User Interface

In order to facilitate automatic user interface generation, services in the OSGi based service framework are decomposed into the following three parts [13]:

- *Program code*: it consists of the service logic and related methods.
- *Interface description*: it provides the template of input/output parameters as well as the device independent presentation of the service. It provides a high-level abstraction of the user interface which takes the form of XML file. Different modalities such as visual, auditory, and tactile are also described in the file.

- *Service profile*: it contains the description of the service properties such as suitability in terms of user group (for blind only), device group (Nokia phone), and functions. The service profile serves mainly for service filtering purpose.

With the service description as input, the UI generator is responsible for automatically generating the graphical user interfaces for different target platforms, taking into the following factors into consideration: interaction techniques, user preferences and device capabilities. The UI generation process is illustrated in Fig. 3.

Aligned with the current trends in web programming, XHTML now only presents structural data of the web pages, while all the layout styling is handled via Cascading Style Sheets (CSS). This dissociation of presentation from model allows us to handle the XHTML form structure separately from its layout.

Utilizing the standard procedure of transforming XML to XHTML, a Service Interface Description (Sid) XML document is pushed through the XSLT processor to produce the XHTML user interface (UI) form. This is the basic version with no stylizing. The use of the XSLT processor and document allows for output to other formats than HTML that can serve as UIs for non-browser-based devices, or web services.

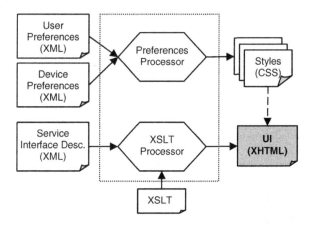

Fig. 3. Generation process of user interfaces according to user context

Our contribution involves providing a Preferences Processor to take user preference and device profile in XML format and generating CSS that would style the output XHTML form appropriately. The system requires that the file for the CSS has a predefined name that the XHTML form knows to point to. Depending on the profiles of the user and device, a different CSS style-sheet would be generated. Since the XHTML file and the Sid XML are not manipulated by the Preferences Processor, developers familiar with the XML-to-XHTML conversion can easily augment their user interfaces with preference-sensitive style-sheets.

By generating the XHTML based on the Sid XML in the server side and providing the CSS based on the user/device profile in the mobile device, the personalized UI can be automatically generated as shown in next section for different users.

5 Implementation of Prototypes

We have implemented a prototype of the impromptu service discovery and access framework using the ProSyst OSGi service platform as the middleware core. All the assistive services are wrapped as OSGi bundles in the framework. Each assistive service implements its specific interface, and each interface has to contain a common method called "getInterfaceDescriptionURL". This method intends to be called by the UI generator when the user selects a service and pass the URL of the service description file to the UI generator. Upon loading the Sid XML file, the UI generator will parse it to generate the corresponding GUI and associate the GUI with the service logic and methods.

In each logical smart space, we install the captive portal on a Linksys WRT54GL WLAN router. On entering the smart space, the web browser of each mobile device is automatically directed to the service portal showing the offered services. For aggregation of devices, we implemented a UPnP wrapper service built over Kono's Cyber-Link for Java UPnP library, providing web access capabilities. Links generated in the service portal point to the presentation URLs of the UPnP services. Fig. 4 shows the system set-up consisting of a Linksys WLAN router, a Samsung Q1 UMPC, an OSGi service platform, an Axis UPnP camera, and a set of assistive services in the shopping mall environment.

For the shopping scenario in a DVD shop, a snapshot of assistive services including a DVD finder and a movie recommender is shown in Fig. 5(a). By choosing the DVD finder service, the user can locate the DVD collections in a certain shelf (Fig. 5(b)). The user can also invoke the movie recommendation service to ask for an impromptu recommendation, based on the user preference and device capability. The detailed recommendation algorithm can be found in [14].

Fig. 4. Impromptu service discovery and access platform

For the lift control use case in the shopping mall, Fig. 6 shows the automatic generated user interfaces of the lift control service for different users and different devices. Depending on the preferences of the user, he would see a different UI. For example, when invoking the lift control service, a young dependent man could see the

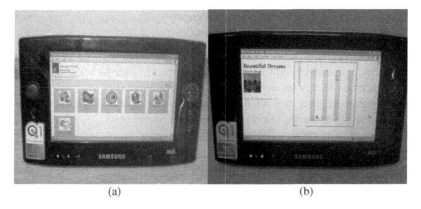

(a)   (b)

Fig. 5. A DVD Shop-specific service: CD finder

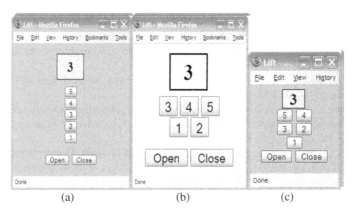

(a)   (b)   (c)

Fig. 6. Generated user interfaces according to user and device context

UI in Fig. 6a with a vivid design. While an elderly person with poor eyesight access the service, the UI generated is shown in Fig. 6b with larger font and a higher contrast ratio. The UI generation is also adaptable to device context such as screen size. Fig. 6c shows the same lift control UI rendered for a smaller screen.

6   Conclusion and Future Work

We have proposed and implemented a framework for impromptu service discovery and access in heterogeneous assistive environments. The distinct features of the framework are:

- Achieving impromptu service discovery with a "minimum assumption" on mobile devices through a proposed captive and service portal mechanism.
- Automatically generating personalized user interface according to the user preference and device capability.

While the framework developed showed encouraging results, there are several issues which we plan to address in the future work:

- Although OSGi service platform can support heterogeneous devices and services, proper tools still need to be developed to transform the other services, especially legacy services into OSGi service bundles.
- When more than one WLAN routers are put in vicinity, the user is still requested to manually select the appropriate access point.
- Security and privacy issues are not addressed in the current system set-up.
- The preferences of different type of dependent users need to be systematically understood and the usability issue of the assistive services and generated user interfaces need to be studied.

References

[1] Edwards, W.K., Newman, M.W., Sedivy, J.Z., Smith, T.F., Izadi, S.: Challenge: Recombinant Computing and the Speakeasy Approach. In: Proceedings of the Eighth ACM International Conference on Mobile Computing and Networking (MobiCom 2002) Atlanta, GA (September 23-28, 2002)
[2] Lee, C., Helal, S., Lee, W.: Universal Interactions with Smart Spaces. IEEE Pervasive Computing 5, 16–21 (2006)
[3] Nakajima, T., Satoh, I.: A software infrastructure for supporting spontaneous and personalized interaction in home computing environments. Personal and Ubiquitous Computing, pp. 379–391 (2006)
[4] Rukzio, E., et al.: An Experimental Comparison of Physical Mobile Interaction Techniques: Touching, Pointing and Scanning. In: Proceedings of the Ubicomp 2006, California, USA (September 17-21, 2006)
[5] Hodes, T.D., Katz, R.H.: Composable ad-hoc location-based services for heterogeneous mobile clients. Wireless Networks, vol. 5(5), pp. 441–427 (1999)
[6] Kindberg, T., et al.: People, Places, Things: Web Presence for the Real World, HPLabs Report, HPL-2001-279 (2001)
[7] Abrams, M.: An Appliance-Independent XML User Interface Language. Computer Networks 31(11-16), 1695–1708 (1999)
[8] Olsen, D.R., et al.: Cross-modal interaction using XWeb. UIST 2000, pp. 191–200 (2000)
[9] Nichols, J., et al.: Generating remote control interfaces for complex appliances. UIST 2002, pp. 161–170 (2002)
[10] Open Service Gateway Initiative (OSGi) http://www.osgi.org
[11] Wang, X.H., Zhang, D.Q., Dong, J., Chin, C.Y., Hettiarachchi, S.R.: Semantic Space: A Semantic Web Infrastructure for Smart Spaces. IEEE Pervasive Computing, vol. 3(2) (2004)
[12] Composite Capabilities/Preferences Profile Home Page http://www.w3.org/Mobile/CCPP/
[13] Ghorbel, M., Renouard, S.: A distributed approach for assistive service provision in pervasive environment. In: Proceedings of the 4th international workshop on Wireless mobile applications and services on WLAN hotspots, pp. 91–100
[14] Yu, Z., Zhou, X.S., Zhang, D.Q., et al.: Supporting Context-Aware Media Recommendations for Smart Phones. IEEE Pervasive Computing 5(3), 68–75 (2006)

Author Index

Lecture Notes in Computer Science

For information about Vols. 1–4439

please contact your bookseller or Springer

Vol. 4493: D. Liu, S. Fei, Z. Hou, H. Zhang, C. Sun (Eds.), Advances in Neural Networks – ISNN 2007, Part III. XXVI, 1215 pages. 2007.

Vol. 4492: D. Liu, S. Fei, Z. Hou, H. Zhang, C. Sun (Eds.), Advances in Neural Networks – ISNN 2007, Part II. XXVII, 1321 pages. 2007.

Vol. 4491: D. Liu, S. Fei, Z.-G. Hou, H. Zhang, C. Sun (Eds.), Advances in Neural Networks – ISNN 2007, Part I. LIV, 1365 pages. 2007.

Vol. 4490: Y. Shi, G.D. van Albada, J. Dongarra, P.M.A. Sloot (Eds.), Computational Science – ICCS 2007, Part IV. XXXVII, 1211 pages. 2007.

Vol. 4489: Y. Shi, G.D. van Albada, J. Dongarra, P.M.A. Sloot (Eds.), Computational Science – ICCS 2007, Part III. XXXVII, 1257 pages. 2007.

Vol. 4488: Y. Shi, G.D. van Albada, J. Dongarra, P.M.A. Sloot (Eds.), Computational Science – ICCS 2007, Part II. XXXV, 1251 pages. 2007.

Vol. 4487: Y. Shi, G.D. van Albada, J. Dongarra, P.M.A. Sloot (Eds.), Computational Science – ICCS 2007, Part I. LXXXI, 1275 pages. 2007.

Vol. 4486: M. Bernardo, J. Hillston (Eds.), Formal Methods for Performance Evaluation. VII, 469 pages. 2007.

Vol. 4485: F. Sgallari, A. Murli, N. Paragios (Eds.), Scale Space and Variational Methods in Computer Vision. XV, 931 pages. 2007.

Vol. 4484: J.-Y. Cai, S.B. Cooper, H. Zhu (Eds.), Theory and Applications of Models of Computation. XIII, 772 pages. 2007.

Vol. 4483: C. Baral, G. Brewka, J. Schlipf (Eds.), Logic Programming and Nonmonotonic Reasoning. IX, 327 pages. 2007. (Sublibrary LNAI).

Vol. 4482: A. An, J. Stefanowski, S. Ramanna, C.J. Butz, W. Pedrycz, G. Wang (Eds.), Rough Sets, Fuzzy Sets, Data Mining and Granular Computing. XIV, 585 pages. 2007. (Sublibrary LNAI).

Vol. 4481: J. Yao, P. Lingras, W.-Z. Wu, M. Szczuka, N.J. Cercone, D. Ślȩzak (Eds.), Rough Sets and Knowledge Technology. XIV, 576 pages. 2007. (Sublibrary LNAI).

Vol. 4480: A. LaMarca, M. Langheinrich, K.N. Truong (Eds.), Pervasive Computing. XIII, 369 pages. 2007.

Vol. 4479: I.F. Akyildiz, R. Sivakumar, E. Ekici, J.C.d. Oliveira, J. McNair (Eds.), NETWORKING 2007. Ad Hoc and Sensor Networks, Wireless Networks, Next Generation Internet. XXVII, 1252 pages. 2007.

Vol. 4478: J. Martí, J.M. Benedí, A.M. Mendonça, J. Serrat (Eds.), Pattern Recognition and Image Analysis, Part II. XXVII, 657 pages. 2007.

Vol. 4477: J. Martí, J.M. Benedí, A.M. Mendonça, J. Serrat (Eds.), Pattern Recognition and Image Analysis, Part I. XXVII, 625 pages. 2007.

Vol. 4476: V. Gorodetsky, C. Zhang, V.A. Skormin, L. Cao (Eds.), Autonomous Intelligent Systems: Multi-Agents and Data Mining. XIII, 323 pages. 2007. (Sublibrary LNAI).

Vol. 4475: P. Crescenzi, G. Prencipe, G. Pucci (Eds.), Fun with Algorithms. X, 273 pages. 2007.

Vol. 4474: G. Prencipe, S. Zaks (Eds.), Structural Information and Communication Complexity. XI, 342 pages. 2007.

Vol. 4472: M. Haindl, J. Kittler, F. Roli (Eds.), Multiple Classifier Systems. XI, 524 pages. 2007.

Vol. 4471: P. Cesar, K. Chorianopoulos, J.F. Jensen (Eds.), Interactive TV: a Shared Experience. XIII, 236 pages. 2007.

Vol. 4470: Q. Wang, D. Pfahl, D.M. Raffo (Eds.), Software Process Dynamics and Agility. XI, 346 pages. 2007.

Vol. 4468: M.M. Bonsangue, E.B. Johnsen (Eds.), Formal Methods for Open Object-Based Distributed Systems. X, 317 pages. 2007.

Vol. 4467: A.L. Murphy, J. Vitek (Eds.), Coordination Models and Languages. X, 325 pages. 2007.

Vol. 4466: F.B. Sachse, G. Seemann (Eds.), Functional Imaging and Modeling of the Heart. XV, 486 pages. 2007.

Vol. 4465: T. Chahed, B. Tuffin (Eds.), Network Control and Optimization. XIII, 305 pages. 2007.

Vol. 4464: E. Dawson, D.S. Wong (Eds.), Information Security Practice and Experience. XIII, 361 pages. 2007.

Vol. 4463: I. Măndoiu, A. Zelikovsky (Eds.), Bioinformatics Research and Applications. XV, 653 pages. 2007. (Sublibrary LNBI).

Vol. 4462: D. Sauveron, K. Markantonakis, A. Bilas, J.-J. Quisquater (Eds.), Information Security Theory and Practices. XII, 255 pages. 2007.

Vol. 4459: C. Cérin, K.-C. Li (Eds.), Advances in Grid and Pervasive Computing. XVI, 759 pages. 2007.

Vol. 4453: T. Speed, H. Huang (Eds.), Research in Computational Molecular Biology. XVI, 550 pages. 2007. (Sublibrary LNBI).

Vol. 4452: M. Fasli, O. Shehory (Eds.), Agent-Mediated Electronic Commerce. VIII, 249 pages. 2007. (Sublibrary LNAI).

Vol. 4451: T.S. Huang, A. Nijholt, M. Pantic, A. Pentland (Eds.), Artifical Intelligence for Human Computing. XVI, 359 pages. 2007. (Sublibrary LNAI).

Vol. 4450: T. Okamoto, X. Wang (Eds.), Public Key Cryptography – PKC 2007. XIII, 491 pages. 2007.

Vol. 4448: M. Giacobini et al. (Ed.), Applications of Evolutionary Computing. XXIII, 755 pages. 2007.

Vol. 4447: E. Marchiori, J.H. Moore, J.C. Rajapakse (Eds.), Evolutionary Computation,Machine Learning and Data Mining in Bioinformatics. XI, 302 pages. 2007.

Vol. 4446: C. Cotta, J. van Hemert (Eds.), Evolutionary Computation in Combinatorial Optimization. XII, 241 pages. 2007.

Vol. 4445: M. Ebner, M. O'Neill, A. Ekárt, L. Vanneschi, A.I. Esparcia-Alcázar (Eds.), Genetic Programming. XI, 382 pages. 2007.

Vol. 4444: T. Reps, M. Sagiv, J. Bauer (Eds.), Program Analysis and Compilation, Theory and Practice. X, 361 pages. 2007.

Vol. 4443: R. Kotagiri, P.R. Krishna, M. Mohania, E. Nantajeewarawat (Eds.), Advances in Databases: Concepts, Systems and Applications. XXI, 1126 pages. 2007.

Vol. 4440: B. Liblit, Cooperative Bug Isolation. XV, 101 pages. 2007.